ALLE ZEIT WACH

1842

Springer Series in Synergetics

Editor: Hermann Haken

Synergetics, an interdisciplinary field of research, is concerned with the cooperation of individual parts of a system that produces macroscopic spatial, temporal or functional structures. It deals with deterministic as well as stochastic processes.

Volume 1 **Synergetics** An Introduction 2nd Edition
By H. Haken

Volume 2 **Synergetics** A Workshop
Editor: H. Haken

Volume 3 **Synergetics** Far from Equilibrium
Editors: A. Pacault and C. Vidal

Volume 4 **Structural Stability in Physics**
Editors: W. Güttinger and H. Eikemeier

Volume 5 **Pattern Formation** by Dynamic Systems and **Pattern Recognition**
Editor: H. Haken

Volume 6 Dynamics of **Synergetic Systems**
Editor: H. Haken

Volume 7 Problems of **Biological Physics**
By L. A. Blumenfeld

Volume 8 **Stochastic Nonlinear Systems** in Physics, Chemistry, and Biology
Editors: L. Arnold and R. Lefever

Volume 9 Numerical Methods in the **Study of Critical Phenomena**
Editors: J. Della Dora, J. Demongeot, and B. Lacolle

Volume 10 **The Kinetic Theory of Electromagnetic Processes**
By Yu. L. Klimontovich

Volume 11 **Chaos and Order in Nature**
Editor: H. Haken

Volume 12 **Nonlinear Phenomena in Chemical Dynamics**
Editors: C. Vidal and A. Pacault

Volume 13 **Handbook of Stochastic Methods** for Physics, Chemistry and the Natural Sciences
By C. W. Gardiner

Volume 14 **Concepts and Models of a Quantitative Sociology**
The Dynamics of Interacting Populations
By W. Weidlich and G. Haag

Volume 15 **Noise Induced Transitions.** Theory and Applications in Physics, Chemistry, and Biology
By W. Horsthemke and R. Lefever

Volume 16 **Physics of Bioenergetic Processes**
By L. A. Blumenfeld

Volume 17 **Evolution of Order and Chaos** in Physics, Chemistry, and Biology
Editor: H. Haken

Volume 18 **The Fokker-Planck-Equation** By H. Risken

W. Weidlich G. Haag

Concepts and Models of a Quantitative Sociology

The Dynamics of Interacting Populations

With 71 Figures

Springer-Verlag Berlin Heidelberg New York 1983

Professor Dr. *Wolfgang Weidlich*
Dr. *Günter Haag*
Institut für Theoretische Physik der Universität Stuttgart, Pfaffenwaldring 57/III
D-7000 Stuttgart 80, Fed. Rep. of Germany

Series Editor
Professor Dr. Hermann Haken

Institut für Theoretische Physik der Universität Stuttgart, Pfaffenwaldring 57/IV
D-7000 Stuttgart 80, Fed. Rep. of Germany

ISBN-13:978-3-642-81791-5 e-ISBN-13:978-3-642-81789-2
DOI: 10.1007978-3-642-81789-2

Library of Congress Cataloging in Publication Data. Weidlich, Wolfgang, 1931- Concepts and models of a quantitative sociology. (Springer series in synergetics ; v. 14) Bibliography: p. Includes index. 1. Sociology-Mathematical models. 2. Equations of motion. 3. Social sciences-Mathematical models. 4. Demography-Mathematical models. 5. System analysis. I. Haag, Günter. II. Title. III. Series. HM24.W448 1982 301'.0724 82-16785

Softcover reprint of the hardcover 1st edition 1983

Typesetting: Schwetzinger Verlagsdruckerei GmbH,

2153/3130-543210

To our parents

Walther and Margarethe Weidlich
Ernst and Elfriede Haag

who conveyed their experience from
more difficult times to their sons

Foreword

While the volumes hitherto published in the Springer Series in Synergetics have been devoted almost exclusively to the self-organized formation of structures in *physics*, *chemistry* and *biology*, the present monograph by *Weidlich* and *Haag* deals with the formation of "structures" (or "patterns") in *society*. At first glance it would seem a daring enterprise to deal with the complex processes in society using concepts and methods first developed in physics. But over the past decade it has been shown that there is a large class of phenomena in a variety of fields to which unifying concepts can be applied. This is particulary true of situations in which a system composed of many parts or individuals acquires a new structure on macroscopic scales. Indeed, this is the definition of synergetics which I formulated more than a decade ago, and which formed the basis of my survey on the profound analogies in the behaviour of complex systems, including those of sociology (H. Haken: *Synergetics. An Introduction*, Volume 1 of this series). As I have pointed out on many occasions, the universal validity of these concepts is neither accidental nor is it caused by a mere extension of physical rules to other fields, but is instead a consequence of deep-rooted structural properties of systems of interacting parts which are due to rigorous mathematical laws.

Generally speaking, concepts and methods originally used in physics can be applied to sociological phenomena in two ways. One approach uses qualitative arguments based on structural analogies. This line was taken in my book *Erfolgsgeheimnisse der Natur; Synergetik: Die Lehre vom Zusammenwirken*, which is presently being translated into English. The other approach is based on detailed mathematical models as initiated by my colleague *W. Weidlich* a decade ago. Over the past years this quantitative approach has been remarkably elaborated by *Weidlich* and his co-workers, who together with *G. Mensch* have also dealt with economic processes.

That a mathematical formulation of social processes produces problems alien to the physical sciences is a fact which cannot be overlooked. A number of these problems are discussed in this volume, and need not be entered into here. Nevertheless, one important problem, that of predictability, should be touched upon. We are often inclined to believe that in physics the course of events can be predicted up to any future time, whereas sociology is sometimes criticized for not being able to do this even for rather short intervals. But over the past years it has become evident in both phase transition theory and synergetics that there are situations in which the macroscopic state of a physical system changes dramatically, but in which it is impossible to predict *which* new macroscopic state will ensue. These situations are a major object of synergetics research, common to all fields ranging from physics to sociology, and they play a central role in this book.

But despite this "*common* drawback" with regard to predictability, there is a fundamental difference between physics and sociology (with biology placed, perhaps, somewhere in between). This difference lies in the question of whether a series of events can be repeated or not. In physics (with the possible exception of astrophysics) we can repeat an experiment over and over again. Thus, even if the outcome of a single experiment is uncertain, by repeating the experiment often enough we can compare the experimental "distribution functions" with the theoretical ones and establish agreement at this level. Such distribution functions are actually well known in sociology, for example in polls. On the other hand, a number of processes in society are of a historical nature, and consist of unique events which can never be repeated. I agree with the authors that even in such cases quantitative models can give us insight into mechanisms of collective phenomena such as the formation of public opinion. Furthermore, ample room remains for field experiments and modelling them on smaller scales, where *distributions* can be directly compared.

Finally, one point which cannot be stressed often enough in the mathematical modelling of sociology is the problem of interpreting model assumptions and model results. It is here that the abstract world of mathematics is filled with "real life". But these problems can be overcome by careful discussion, and I am convinced that the mathematical approach will help in establishing which assumptions are consciously or subconsciously made in dealing with certain sociological problems.

I am sure that the present book will be highly stimulating and useful to students and researchers in sociology and possibly in other fields as well.

Stuttgart, Autumn 1982 *H. Haken*

Preface

Remarkable progress has been made during the last years in the natural sciences towards the development of interdisciplinary unifying concepts which include principles ranging from those of the traditional disciplines of physics and chemistry to those of biology and which can be seen as focussing in synergetics. The authors – after having published some mathematical models of sociological processes – felt challenged by this development to give a coherent and general description of some new concepts on quantitative sociology which correspond to these interdisciplinary ideas. The scope of this book is determined by this objective.

Transferring the mathematical methods of the natural sciences to the social sciences (including economics) involves considering the general problems of structural analogies and differences between them. Hence, the interpretation of and the limitations to be imposed on the quantification of sociological processes has to be considered at some length (Capters 1 and 6). While trying to achieve this transfer it was found that some mathematical methods of statistical physics and synergetics prove to be applicable to sociology. On the other hand, new concepts taking account of the characteristic properties of social systems had to be introduced.

The book is organized as follows:
In the introductory chapter the unifying concepts applicable to natural and social sciences are described qualitatively. Chapter 2 treats a simple example of quantitative sociology, the "opinion formation model" – the mathematical tools needed are presented in a self-contained manner. Chapter 3 contains some general concepts of quantitative sociology whereby a generalization of those introduced in Chapter 2 is achieved.

The second part of the book, consisting of Chapters 4, 5 and 6, is devoted to applications which are intended firstly to be neither trivial nor of such an overwhelming complexity as to disguise the underlying structures – it was not so easy to find this kind of model – and secondly to belong to relevant and differing fields within the social sciences.

Chapter 4 is concerned with population dynamics in the more traditional sense, including birth, death and migration processes. However, the approach includes migration between distant places as well as nonlinear effects. Hence it goes further than most treatments of this dynamics and also seems to be suitable for modelling migration processes in the human context. Further special cases of the theory are Volterra-Lotka processes combined with migration. Chapter 5 treats non-equilibrium investment fluctuations of the Schumpeter type. The model describes the development of behaviour in the decision-making of investors in interaction with economic variables. Finally, Chapter 6 considers the problem of interacting macrosocieties. In comparison to other "world models" the "minimal model" presented is rather simple but does include macrovariables of political psychology and their interaction with material aspects of the societies. Some of the more technical sections concerning methods and relations of the mathematical apparatus needed are marked by an asterisk and may be skipped in a first reading.

The authors wish to thank Dr. *H. Lotsch* of Springer-Verlag for arranging the publication of our book in the "Springer Series in Synergetics".

We would like to thank Prof. *H. Haken*, the inaugurator of synergetics and a friend and colleague of one of the authors (W. W.) for his encouragement in our writing this book and for his many suggestions in fruitful discussions that have taken place over the years.

Thanks are also due to the friend and mentor of our Institute of Theoretical Physics, Prof. *H. Fröhlich*, F.R.S., who very early supported the idea of integrating sociological phenomena into a quantitative framework.

While writing the manuscript we were introduced to non-equilibrium problems of economics by Prof. *G. Mensch*, to whom we are most grateful for his ensuing close cooperation on the "Schumpeter Clock" model presented in Chapter 5.

Finally, we thank the friends and colleagues of our institute, Professors *M. Wagner*, *G. Mahler*, *U. Weiss* and *H. Grabert*, for many valuable discussions.

Furthermore we are indebted to Dr. *H. G. E. Hentschel* for critically reading parts of the manuscript and for making many valuable suggestions. In particular, we would like to thank Dr. *C. Gee* who thoroughly revised the whole manuscript and made many substantial proposals for its improvement.

Last, but not least, special thanks go to our secretary, Miss *Eva Effenberg*, who typed the many different versions of the manuscript with great patience and care.

Stuttgart, June 1982 *Wolfgang Weidlich · Günter Haag*

Contents

Sections marked with * can be omitted on a first reading

1. Introduction and Outline

We are going to develop a framework for the quantitative description of the dynamics of a wide class of sociological phenomena. The concepts of the theory used belong to the field of synergetics, a new interdisciplinary branch of science.

In this first chapter an introductory and qualitative account is given of the non-trivial problem of why and to what extent it is possible to describe dynamic phenomena of a society in a quantitative or at least semi-quantitative manner. Three steps are involved:

Firstly, some fundamental concepts of synergetics originally developed for physical and chemical systems are explained.

Secondly the question is considered for which class of phenomena in a society the transfer of synergetic concepts is appropriate and where these concepts are of limited value only.

Thirdly, the relation between the traditional forms of sociology and the quantitative approach adopted here is discussed.

1.1 Synergetic Concepts in the Natural Sciences

Synergetics is defined as the science of collective static or dynamic phenomena in closed or open multi-component systems with "cooperative" interactions occurring between the units of the system [1.1–9].

In physics, chemistry and biology synergetics concentrates on the structural self-organizing space-time features of systems on a macroscopic level [1.10–23]. It turns out that on this level there exist close analogies between various systems, although they are composed of different units with completely different elementary interactions. Due to this fact the concepts of synergetics are of interdisciplinary universality. The word "synergetics" and many of the unifying concepts have been introduced into the natural sciences by *Haken* [1.1, 3, 10, 12].

Before trying to transfer such concepts to sociology it is worthwhile to appraise and summarize their meaning for physico-chemical systems, where they have already proved to be of value.

1.1.1 Physico-Chemical Systems

In *physics and chemistry* the *elementary units* composing the basic system are well known. These consist of elementary particles and photons or of composite units like atoms and molecules. The *interactions* between the units are also known; there exist strong, electromagnetic, weak and gravitational interactions between elementary particles. More complex interactions like the interatomic forces responsible for chemical bonds and the *van der Waals* forces between molecules of a gas can be derived from these elementary interactions. The elementary units, their position and momentum vectors and their interactions constitute the basic microscopic level of a physico-chemical system.

On the other hand, the "gross" features of a system consisting of many particles are relatively easily accessible to experiment on the macroscopic level.

According to experimental conditions the system may be *closed* or *open*. A *closed system* is at most confined by static boundary conditions. It has *internal interactions*, but no further interaction with the environment. An *open system*, however, has not only internal interactions between its constituents, but also *external interactions* with the environment, for instance, with external heat and particle reservoirs by exchange of energy and particles.

Closed or open systems on the macroscopic level turn out to be characterized by relatively few *collective macroscopic observable quantities*, also denoted here as *grossvariables* or *macrovariables*. Examples are: pressure, density, entropy, energy- and particle-fluxes, correlation functions and further parameters characterizing the macroscopic state or dynamic space-time structure of fields and particles.

If the external conditions of an open system are changed by varying certain *control parameters* the system may undergo a *radical change* in its macroscopic global state if the control parameters pass certain critical values. Such transitions are denoted as *phase transitions* [1.10, 11, 13, 14, 17, 24, 48].

Since in a phase transition the macroscopic order changes in essential respects, it is important to find suitable macrovariables characterizing this change of order. Such macrovariables are denoted as *order parameters*. Therefore, in a phase transition new order parameters arise or, vice versa, old order parameters decay.

Two examples of phase transitions illustrate this point:

a) One observes the separation of a homogeneous gas in a given volume into a gaseous state and a liquid state by lowering the temperature, i.e. the control parameter, below a critical value T_c. The suitable order parameter characterizing this macroscopic change of order is the newly arising difference of density between the liquid and gaseous part of the system at temperatures below T_c [1.25, 3.8–11].

b) In a laser the generation of a coherent light beam is observed if one increases the pumping of atoms into excited states beyond a critical threshold value p_c. Here, the pumping intensity p is the control parameter and the newly arising intensity of the laser light is the order parameter characterizing the phase of laser action [1.11, 12, 26, 27, 28].

Case a) illustrates a phase transition in thermal equilibrium and b) a dynamic phase transition far removed from the state of thermal equilibrium.

If the control parameter approaches the vicinity of a critical value for a phase transition, the system undergoes *critical fluctuations*. These fluctuations, which are of considerable magnitude, are *necessarily connected* with the phase transition and in a sense *anticipate* the rise or decay of the order parameters.

For example, in case a) above it is found that for $T \gtrsim T_c$ critical density fluctuations of the still homogeneous gas occur, which anticipate the separation into a gaseous and a liquid state. These density fluctuations can be measured as they scatter light passing through the gas and yield the phenomenon of critical opalescence. In the case b) at levels $p \lesssim p_c$ slightly below threshold, i.e. before laser action sets in, large photon intensity fluctuations are found to occur preparatory to the creation of the steady laser beam obtained in the new phase for $p > p_c$.

1.1.2 Dynamics and Equations of Motion in Physico-Chemical Systems

It is now one of the fundamental problems of statistical physics and synergetics to derive the properties of physico-chemical multi-component systems on the macroscopic level from their constituent components on the elementary microscopic level. Part of the problem is to explain *which macrovariables* (including order parameters) may be relevant under given circumstances and to describe their dynamics by appropriate *equations of motion* [1.49, 50].

General concepts have been developed to establish relevant macrovariables and to derive formally exact equations of motion for the macrovariables from a fundamental standpoint [1.29, 30].

On the other hand, *phenomenological derivations* of such equations are possible by making some approximations and plausible assumptions about the system. Simple but intuitively plausible forms for these equations will now be discussed. Exact equations of motion for macrovariables contain the interactions between the macrovariables as well as the rapidly fluctuating random forces due to the influence of the microvariables [1.31–33, 51–53].

The simplest type of such equations for macrovariables either derived phenomenologically or from fundamental dynamics is that of the *"Langevin equations"*, a set of first-order differential equations in the time variable t

$$\frac{dx_i(t)}{dt} = F_i(x_1, \ldots, x_n) + f_i, \qquad \text{with} \quad i = 1, 2, \ldots, n \tag{1.1}$$

where f_i are the fluctuating random forces and $F_i(x_1, \ldots, x_n)$ are – in general non-linear – functions of the macrovariables $x_j(t)$ [1.1, 3, 34].

This description is illustrated typically by writing down the Langevin equations for *Brownian motion*:

$$\frac{dx_i}{dt} = \frac{1}{m} p_i \qquad \text{and} \quad \frac{dp_i}{dt} = -\gamma p_i + f_i \qquad \text{with} \quad i = 1, 2, 3 \, . \tag{1.2}$$

Here, x_i and p_i are the position and momentum variables of a small "Brownian" (dust) particle suspended in a gas. The motion of this particle is influenced by collisions with the surrounding molecules. The force exerted on the Brownian particle is split into a "systematic" part $-\gamma p_i$ and a "random" fluctuating component f_i. Neglecting the f_i, the Brownian particle performs a damped motion and finally comes to rest. The effect of f_i, however, leads to a never ending irregular motion of the particle, which can be easily observed under the microscope (Fig. 1.1).

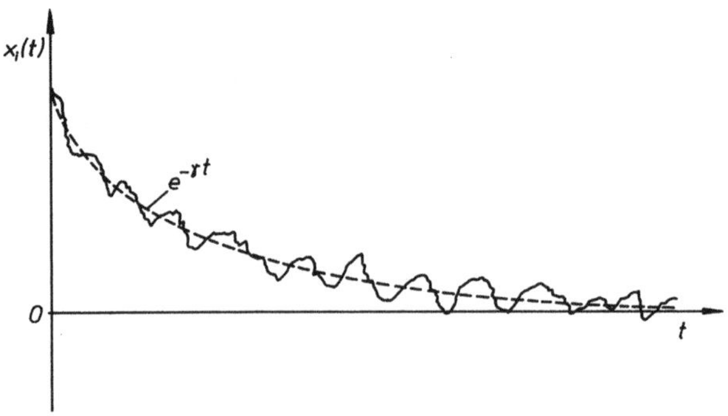

Fig. 1.1. The path $x_i(t)$ of a Brownian particle performing a systematic motion superimposed by random fluctuations

For the following general discussion of the physical meaning of the Langevin equations the so-called *ensemble standpoint* is introduced: An ensemble of macrosystems with equal composition and described by the same set of macrovariables x_i is considered. The members α of this ensemble, however, may differ with respect to the microscopic fluctuating random forces f_i which usually are not under experimental control. Although starting with the same initial conditions $x_i^\alpha(0)$ for the macrovariables, different paths $x_i^\alpha(t)$ for the members of the ensemble will be obtained because of the influence of the differing fluctuating forces in (1.1). We denote the *ensemble average* or *ensemble mean value* of a – in general time dependent – quantity $x^\alpha(t)$ by $\langle x(t)\rangle$. If the ensemble consists of samples $\alpha = 1, 2, \ldots, \omega$ with paths $x_i^\alpha(t)$, the ensemble mean value is defined by

$$\langle x_i(t)\rangle = \frac{1}{\omega}\sum_{\alpha=1}^{\omega} x_i^\alpha(t) \tag{1.3}$$

Returning to the fluctuating random forces, due to the lack of full information about the microscopic state of the system it is only possible to derive (or to postulate phenomenologically) certain *ensemble properties* of these forces expressed by ensemble averages.

For a wide class of systems these random forces obey the following equations [1.1]

$$f_i(t) = \sum_{j=1}^{n} g_{ij}(x_1(t), \ldots, x_n(t))\, \xi_j(t)\ . \tag{1.4}$$

Equation (1.4) implies that $f_i(t)$ consists of a sum of random forces $\xi_j(t)$ with strengths factors $g_{ij}(x_1, \dots, x_n)$ in general dependent on the macrovariables. The random forces $\xi_j(t)$ have vanishing ensemble meanvalues

$$\langle \xi_j(t) \rangle = 0 \,. \tag{1.5}$$

Further, they have no correlation with the macrovariables at the same or at previous points in time

$$\langle \xi_j(t+\tau)\, G(x_1(t), \dots, x_n(t)) \rangle = 0 \qquad \text{for} \quad \tau \geq 0 \tag{1.6}$$

from which it follows that

$$\langle f_i(t) \rangle = 0 \,. \tag{1.7}$$

The random forces $\xi_j(t)$ are, however, correlated with themselves over very short time periods

$$\langle \xi_j(t+\tau)\, \xi_i(t) \rangle = \delta_{ij}\, \delta(\tau) \tag{1.8}$$

where $\delta(\tau)$ is the so-called delta function.

In order to discuss which forms of solutions to (1.1) can be expected, two structurally different cases have to be distinguished:

a) Suppose that the solutions of (1.1) starting from the same initial values $x_i^\alpha(0)$, i.e. the single sample paths $x_i^\alpha(t)$, do not deviate essentially from their mean values $\langle x_i(t) \rangle$ during the whole time development. In this case the variances (mean square deviations) defined by

$$\begin{aligned} \sigma_i(t) &= \langle (x_i(t) - \langle x_i(t) \rangle)^2 \rangle \\ &\equiv \langle x_i^2(t) \rangle - \langle x_i(t) \rangle^2 \end{aligned} \tag{1.9}$$

of the ensemble can be presumed to remain small at all points in time. After transition to mean values in (1.1) the replacement

$$\langle F_i(x_1(t), \dots, x_n(t)) \rangle \approx F_1(\langle x_1(t) \rangle, \dots, \langle x_n(t) \rangle) \tag{1.10}$$

and (1.7) lead to the approximate equations for the mean values $\langle x_i(t) \rangle$ of $x_i^\alpha(t)$:

$$\frac{d\langle x_i(t) \rangle}{dt} \approx F_i(\langle x_1(t) \rangle, \dots, \langle x_n(t) \rangle) \qquad \text{with} \quad i = 1, 2 \dots, n \,. \tag{1.11}$$

The Eqs. (1.11) are an autonomous closed non-linear set of differential equations for the mean values $\langle x_i(t) \rangle$ which, by assumption, do not deviate essentially from the true sample paths $x_i^\alpha(t)$. Hence, in this case the macrovariables $x_i^\alpha(t)$ approximately obey a *closed self-contained subdynamics*. The intui-

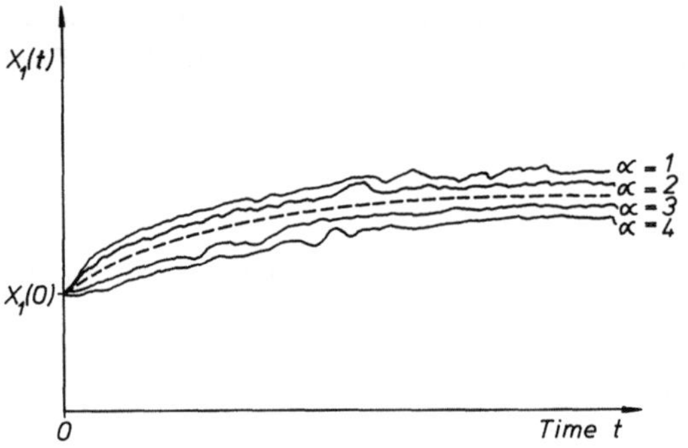

Fig. 1.2. Behaviour of one representative macrovariable x_1 in the case a) of a quasi-closed subdynamics for macrovariables. (—) Selected paths $x_1^\alpha(t)$ of the ensemble. (---) Ensemble mean value $\langle x_1(t)\rangle$ of the $x_1^\alpha(t)$

tive reason for this behaviour is that the quickly fluctuating random forces exerted by microvariables only lead to small fluctuations of the macrovariables $x_i^\alpha(t)$ around their smoothly developing ensemble average $\langle x_i(t)\rangle$ (Fig. 1.2).

b) In contrast to case a) it may happen that the equation

$$\frac{dx_i(t)}{dt} = F_i(x_1, \ldots, x_n) \qquad \text{with} \quad i = 1, 2 \ldots, n \tag{1.11 a}$$

obtained from (1.1) by neglecting the fluctuating forces f_i shows *"bifurcation behaviour"* [1.1, 2, 5, 10, 54, 55]. This means that solutions of (1.11 a) starting with slightly different initial values $x_i(0)$ in the vicinity of a *"critical value"* x_{ic} (e.g. an unstable stationary point) have completely different paths $x_i(t)$. Thus, adding fluctuating forces f_i to the rhs of (1.11 a) and therefore returning to the form (1.1), completely different ("bifurcated") paths for different members of the ensemble can be obtained. In other words, infinitesimally small differences of certain "causes", i.e. the random forces, may lead to very large differences of the "effects", i.e. the paths of the macrovariables.

The individual paths $x_i^\alpha(t)$ then deviate significantly from their mean value $\langle x_i(t)\rangle$. Therefore (1.10) is no longer generally valid and the subdynamics for $\langle x_i(t)\rangle$ is not characterised by (1.11). Further, the mean values $\langle x_i(t)\rangle$ are no longer characteristic for the sample paths $x_i^\alpha(t)$ (Figs. 1.3, 1.4).

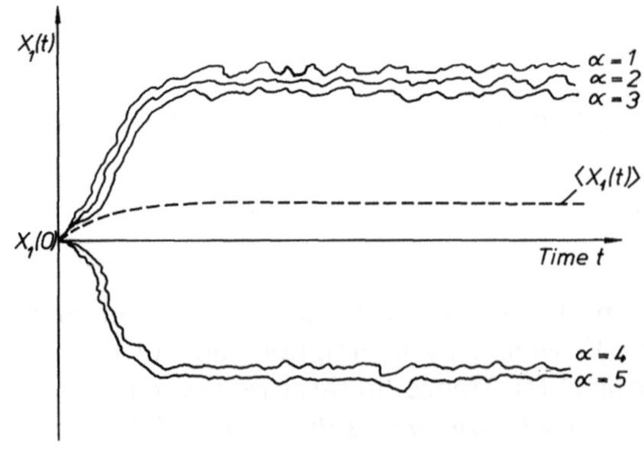

Fig. 1.3. Behaviour of one representative macrovariable x_1 in the case b) of a simple bifurcation, the development into two stable modes. (—) selected paths $x_1^\alpha(t)$ of the ensemble. (---) ensemble mean value $\langle x_1(t)\rangle$ of the $x_1^\alpha(t)$

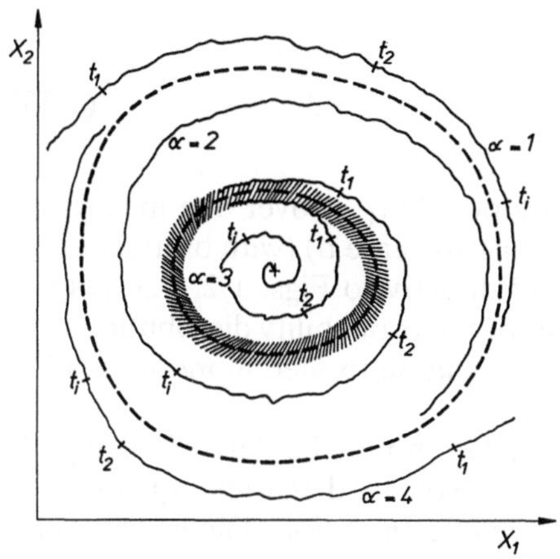

Fig. 1.4. Motion of two representative macrovariables $x_1(t)$, $x_2(t)$ in the case b) for an example where a stable focus, an unstable limit cycle (separatrix) and a stable limit cycle (---) exist. Representative paths $\{x_1^\alpha(t), x_2^\alpha(t)\}$ (—) are plotted in the x_1/x_2 plane. In the critical (//////) domain around the separatrix small fluctuations may trigger the bifurcation between subsequently diverging paths. The time t ($t_1 < t_2 < \cdots t_i$) parametrizes the paths

The cases a) and b) lead to very different structural consequences.

In case a) the already mentioned quasi-closed self-contained quasi-deterministic subdynamics of macrovariables is obtained.

In case b) small and unpredictable microfluctuations may decisively influence the macrovariables in a critical domain and finally trigger their bifurcation, i.e. their development into globally different macrostates or phases of the system. This means that in case b) the system undergoes phase transitions as described above.

A second *probabilistic description* of systems on a macroscopic level is now introduced which takes into account from the very beginning the fact that the exact values of macrovariables are unknown because of the fluctuating random influence of microvariables.

Starting from the ensemble standpoint, the probability

$$P(x_1, \ldots, x_n; t)\, d\tau(\boldsymbol{x}) \qquad \text{where} \quad d\tau(\boldsymbol{x}) \equiv dx_1 dx_2 \ldots, dx_n \tag{1.12}$$

of finding the macrovariables $x_1, \ldots, x_n$ in the interval $d\tau(\boldsymbol{x})$ between $\{x_j\}$ and $\{x_j + dx_j\}$ at time t is introduced. It will be shown in Chap. 3 that the set (1.1) of Langevin equations with fluctuating forces obeying (1.4–8) is completely equivalent to the following *Fokker-Planck equation* of motion for the probability distribution function

$$\frac{\partial P(\boldsymbol{x}; t)}{\partial t} = -\sum_{i=1}^{n} \frac{\partial}{\partial x_i} [K_i(\boldsymbol{x}) P(\boldsymbol{x}; t)] + \frac{1}{2} \sum_{i,j=1}^{n} \frac{\partial^2}{\partial x_i \partial x_j} [Q_{ij}(\boldsymbol{x}) P(\boldsymbol{x}; t)] \tag{1.13}$$

with

$$K_i(\boldsymbol{x}) = F_i(\boldsymbol{x}) + \frac{1}{2} \sum_{l,j=1}^{n} \frac{\partial g_{ij}(\boldsymbol{x})}{\partial x_l}\, g_{lj}(\boldsymbol{x}) \tag{1.14}$$

and

$$Q_{ij}(\boldsymbol{x}) = \sum_{l=1}^{n} g_{il}(\boldsymbol{x}) \cdot g_{jl}(\boldsymbol{x}) . \tag{1.15}$$

The development of a probability distribution $P(x_1; t)$ over one macrovariable x_1 with time in case a) (without bifurcation) and case b) (with bifurcation) is illustrated in Figs. 1.5, 1.6 which are related directly to Figs. 1.2, 1.3, respectively, describing the same dynamics in terms of the probability distribution over the ensemble $x_1^a(t)$. This one dimensional case will be treated in more detail in Chap. 2.

On the other hand, the Fokker-Planck equation may be derived (Chap. 3) using some approximation procedures from a so-called master equation for $P(\boldsymbol{x}; t)$ which has a particularly simple intuitive interpretation. The *master equation* has the form

$$\frac{\partial P(\boldsymbol{y}; t)}{\partial t} = \int_{(\boldsymbol{x})} d\tau(\boldsymbol{x})[w(\boldsymbol{y} \leftarrow \boldsymbol{x}) P(\boldsymbol{x}; t) - w(\boldsymbol{x} \leftarrow \boldsymbol{y}) P(\boldsymbol{y}; t)] . \tag{1.16}$$

Here, $d\tau(\boldsymbol{x})\, w(\boldsymbol{x} \leftarrow \boldsymbol{y})$ is the (positive definite) transition probability per unit time from values $\{y_k\}$ of the macrovariables into the cell $d\tau(\boldsymbol{x})$ of the configuration space.

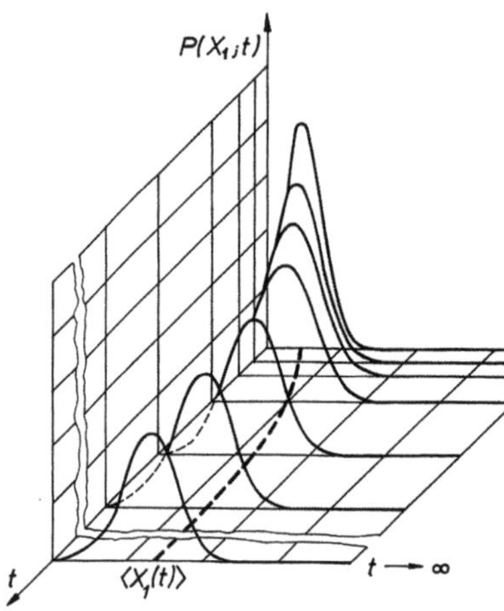

Fig. 1.5. Evolution of a probability distribution $P(x_1; t)$ and of the corresponding mean value $\langle x_1(t)\rangle$ (---). Case a)

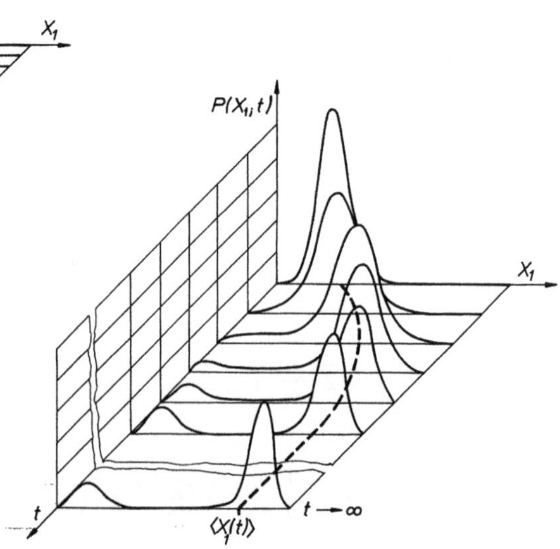

Fig. 1.6. Evolution of a probability distribution $P(x_1; t)$ and of the corresponding mean value $\langle x_1(t)\rangle$ (---). Case b)

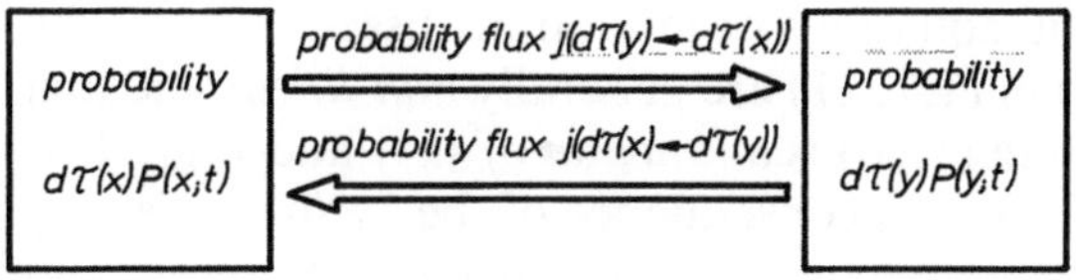

Fig. 1.7. Probability fluxes $j(d\tau(y) \leftarrow d\tau(x))$ and $j(d\tau(x) \leftarrow d\tau(y))$ between two representative cells of the configuration space

The meaning of the master equation (1.16) then becomes clear from probability balance considerations: The change in time $[\partial P(y;\,t)/\partial t]\,d\tau(y)$ of the probability in the interval $d\tau(y)$ is due to the probability flux

$$\int_{(x)} j[d\tau(y) \leftarrow d\tau(x)] \equiv \int_{(x)} [d\tau(y)\,w(y \leftarrow x)]\,d\tau(x)\,P(x;\,t)$$

from all points $\{x_k\}$ *into* the cell $d\tau(y)$ [first term on the rhs of (1.16)] as well as to the probability flux

$$\int_{(x)} j[d\tau(x) \leftarrow d\tau(y)] \equiv \int_{(x)} [d\tau(x)\,w(x \leftarrow y)]\,d\tau(y)\,P(y;\,t)$$

out of the cell $d\tau(y)$ *to* all points $\{x_k\}$ [second term on the rhs of (1.16)]. The probability flux situation between two representative cells $d\tau(x)$ and $d\tau(y)$ in macrovariable space is symbolically illustrated in Fig. 1.7.

Both equations, the master equation and the Fokker-Planck equation, have one important property: An arbitrary initial distribution – for instance, concentrated in a small interval $d\tau(x)$ – finally develops into a *stationary* equilibrium distribution $P_{st}(x)$.

Some conclusions obtained by comparing this result with the (equivalent) Langevin equation picture are:

a) The ensemble of Langevin paths spreading out from any initial cell $d\tau(x)$ will finally have to reach the whole range covered by the stationary probability distribution.

b) The stationary distribution also describes an ensemble of points, each of them traversing a Langevin path, which can reach every cell $d\tau(y)$ of the range of $P_{st}(x)$ in the course of time. On the other hand, $P_{st}(x)$ does not change its shape with time. These two facts can only be reconciled if, for a very long time interval $T \to \infty$, every Langevin path spends a fraction of time proportional to $P_{st}(y)\,d\tau(y)$ in each interval $d\tau(y)$. This leads to the conclusion that the *ensemble mean value* of any function $F(x)$ in the stationary ensemble can be identified with its *time mean value* taken along a Langevin path $x^{\alpha}(t)$:

$$\begin{aligned}\langle F(x)\rangle_{st} &= \int_{(x)} F(x)\,P_{st}(x)\,d\tau(x)\\ &= \lim_{T\to\infty} \frac{1}{T}\int_0^T F(x^{\alpha}(t))\,dt\,. \end{aligned} \tag{1.17}$$

A further essential feature of the equations of motion discussed so far is the fact that the functions $F_i(x_1, \ldots, x_n)$ in (1.1, 11) are generally *non-linear* and that the transition probability $w(y \leftarrow x)$ in (1.16) is a *function of* $\{x_k\}$ and $\{y_k\}$. This non-linear structure can be considered as a consequence of the *"self-consistency principle"*, i.e. there exists a cyclic coupling or a feedback between "causes" and "effects", where both may be macrovariables. For instance, the units of a system may generate a collective field; this field reacts on the units in a static or dynamic way creating and stabilizing a certain space-time order of positions, orbits, excitations, etc. of the units. The ordered structures and the collective field then mutually sustain themselves in a self-consistent way.

The type of equations that apply to the dynamics of physico-chemical macrosystems have been described at some length, because the equations of quantitative sociology to be introduced in the forthcoming chapters have the same general structure.

1.2 Synergetic Concepts in Sociology

Before trying to apply synergetic concepts to sociology it must be stated that there exists no direct "short-cut" to transfer physico-chemical concepts to sociology. Such a direct transfer could at best lead to a superficial analogy between both sciences. It is necessary, firstly, to develop concepts appropriate and characteristic for the quantitative description of sociological processes. Secondly, it has to be examined whether – or under which circumstances – synergetics can be applied to genuine sociological phenomena. If this turns out to be possible a *structural relationship* between the natural and social sciences, and not only an accidental analogy will have been found.

In recent years some authors working e.g. on physico-chemical open multicomponent systems have become aware of the applicability of the more general concepts of synergetics and of quantitative methods such as those discussed in Sect. 1.1 to the investigation of the structure of sociological systems and of the historical development of societies [1.35–47].

Evidently, the human society is a multi-component system composed of "units", namely its individual members. This system is *open* because there exists not only an *internal* interaction of a material nature and through ideas between the members of a society but also an interaction with the *external* environment and the technological surroundings.

Starting from this fact, the possibility of a transfer of synergetic concepts from physico-chemical systems to sociological systems is based on the following general line of arguement.

Firstly a physico-chemical system is assumed as being composed of a large number of units, each of them existing in one of several different possible states. Then it is assumed that the dynamics of the system consists – on the microscopic level – of correlated transitions of the units between the possible states. After introducing appropriate collective variables describing the global motion of the system, these variables will satisfy equations of motion of the type described in Sect. 1.1.2.

Secondly a society can also be regarded as being composed of a large number of members, who individually adopt different attitudes or "states" of behaviour. Furthermore it can be assumed that the causal changes in the society are correlated changes in the attitudes of its members. The global change of behaviour can then be obtained by introducing appropriate grossvariables, for instance, in terms of average attitudes of groups of members of the society.

If a *formal isomorphism* between the "natural" and the sociological type of system can be established, such that

1) the units of the systems correspond,
2) the transitions between the states of the units in one system correspond to those in the other system,
3) the grossvariables introduced in one system can be uniquely assigned to those in the other system,

then it is reasonably obvious that the dynamics of both systems – as far as they are described by the transitions postulated – is formally identical and independent of the nature of the units or individuals composing them. Even in less exacting situations of comparison, where the close formal correspondence of the sociological system to a non-sociological system cannot be uniquely established, it is intuitively understandable that individual transitions between attitudes lead to the kind of global dynamics described by equations of the type discussed in Sect. 1.1.2. The formal theory of such sociological processes will be established on this basis in Chap. 3.

On the other hand, and in contrast to physico-chemical systems, however, the "units" and the "interactions" of a human society are not of an elementary but of a highly complex nature: In many cases the thoughts, emotions and intentions of a single individual may exhibit a complexity reflecting the complex structure of the brain itself. Furthermore, the interactions between individuals do not possess a definite form, since, depending on the existing situation, the interactions are either activated and present or are deactivated and latent. Another fundamental difference is that in physico-chemical systems the interactions lead to a "blind", i.e. unintended self-organization of the system, whereas the nature of interactions in sociology leads, at least partially, to intelligent "planned" structures.

In order to have a useful and therefore still sufficiently simple quantitative theory, those sociologic phenomena have to be *excluded* from consideration for which the full complexity of human behaviour and interactions has to be taken into account in the sense of a comprehensive *"microscopic theory"*.

Two examples of social phenomena, which – for different reasons – *cannot* or which are included only to a certain extent in the approach of this book are:

a) The behaviour of partners in a marriage, where the highly complex psychology of personal preferences is activated and where no ensemble mean value smoothes out the complexity. An elaborate theory of brain structure sufficiently definite to allow of a quantitative treatment of such behaviour in detail is far from being available.

b) *"Normative economy"*, i.e. the theory of intelligent and intentional optimization of decisions to create well organized structures in the economy and other fields of applied management with respect to given aims and constraints is only of marginal interest here. The problems of this science can be treated mathematically, e.g. in terms of game theory.

However, the results of economic decisions can and will be included in the approach here but only in as far as they are *realized* and have an *impact* on real sociological processes.

The class of sociological phenomena which are the main subjects of the quantitative approach of this book will now be defined. These phenomena belong to the field of *socio-political psychology* of *individuals* on the *"microlevel"* and the consequential *collective material, economic and abstract structures* on the *"macrolevel"* of *society*.

Under discussion are those phenomena where – on the level of the single person – a complex mixture of fluctuating rational considerations, professional activities and emotional preferences and motivations *finally merge* into one of relatively few well demarcated *resultant attitudes*. These attitudes may be in the field (later denoted as "dimension") of religion, politics, education, habitation, occupation, economic standard, consumer habits, family, sport, etc. A single "attitude" may, for instance, find an expression in the form of a religious denomination, in a vote for a certain political party or in membership of a certain income class.

The same fluctuating microthoughts, emotions and experiences finally merging into an attitude may from time to time also lead to a *transition* from one attitude to another. Further, even new, hitherto unknown attitudes may develop in a sociological process; hence, the *manifold* of possible attitudes is an *open* one. In Chap. 3 this manifold is described in terms of an *"attitude space"*.

In effecting transfers from the microlevel of individuals to the macrolevel of society the concept of *"socio-configuration"* will be introduced. The socio-configuration describes the distribution of attitudes over the subpopulations of a society and may be considered as an appropriate *set of macrovariables* for this society.

The *"dynamics of interacting populations"* to be investigated in this book then consist of a theory concerning the structural development of this socio-configuration, or of quantities derived from it, with time.

The procedure in setting up this theory is based on a *formal* structural similarity between statistical physics and sociology, as explained above. This similarity, however, is *not at all* due to any (misleading) *direct* similarity between units, interactions and variables of physical and sociological systems, but is a consequence of the *general probabilistic laws* governing the statistical dynamics of multi-component systems!

In sociology the "microscopic" details of the complex interplay of rational and emotional, conscious and subconscious, genetic and environmental influences on individual decisions are unknown. It has been seen, (1.13, 16), that in physics a *probabilistic description* of the motion of macrovariables proves to be adequate even when the details of the microfluctuations of the system are *unknown*. Therefore a probabilistic description of decision and motion processes in a society could also prove to be adequate. As a consequence the concept of a *transition probability* for a member of a subpopulation to change from one attitude to another will be introduced in Chap. 3. Beginning with this concept equations of motion for the socio-configuration will be derived which structurally resemble the form of (1.16) of systems in the natural sciences.

It has been mentioned already that in physics such equations may often be derived from *first principles*. In sociology, however, resort has to be made to a *phenomenological* approach starting from appropriate but flexible choices of the socio-configuration and the transition probabilities. As it is wished to include both variables of the economy and of political psychology, the background observations and laws for this phenomenological approach have to be taken from *economics* and *politics*. This will become explicit in the discussion of the models of the following chapters.

It is, even at this point, possible to anticipate some general consequences of the proposed quantitative description of society which can be understood qualitatively and intuitively without explicit use of the mathematical tools developed later.

As in physics, the *non-linear form* of the equations expresses a *self-consistent structure* of the sociological system. This implies that there exists a *cyclic coupling* between causes and effects in society which can be described as follows:

The individual members of society contribute through their cultural and economic activities to the generation of a general "field" of civilisation with cultural, political, religious, social and economic components. This collective field determines the socio-political atmosphere and the cultural and economic standard of the society and may be considered as an order parameter of the system characterizing the *phase* in which the society exists.

Vice versa, the collective field strongly influences the individuals in the society by orientating their activities, by activating or deactivating their latent positive and negative qualities and capabilities and by extending or narrowing their scope of thinking and action.

One of the features of this sort of cyclic coupling of causes and effects is that self-accelerating as well as self-saturating processes result. In the "normal case" the feedback between individuals and the collective field leads to a sustained quasi-stable or evolutionary phase, i.e. a "self-consistent" collective structure of society. In this case there exists a certain predictability of further development because the space of mutually influencing relevant macrovariables (taken from the socio-political and economic situation) is known and these macrovariables obey a *quasi-closed subdynamics*.

If, however, the control parameters governing the dynamic behaviour of the system attain certain *critical values* due to internal or external interactions, the macrovariables may move into a critical domain out of which highly divergent alternative paths are possible. In this situation small unpredictable *microfluctuations* – for instance, the actions of very few influential people – may decide into which of the diverging paths the behaviour of the society will *bifurcate*. The branching into different possible scenarios at the bifurcation point may be accompanied or promoted by the *rise in significance of new, hitherto unknown or irrelevant order parameters*. For instance, a new ideology or a new scientific discovery may gain decisive influence. In this case the attitude space and the socio-configuration may have to be extended by including new dimensions in order to model the dynamics of the new macrovariables. The nessecary extension of the framework of description implies that the original set of socio-political macrovariables is no longer sufficient and no longer obeys a closed subdynamics; the *predictability* with respect to the old set of macrovariables has been lost.

In physics this bifurcation process is called a phase transition. With respect to a society one may refer to a *revolutionary situation* in which e.g. a few resolute individuals may trigger a *revolution*, i.e. a sudden transition into a new global "phase" of the society. The transition from one phase to another shows that self-consistent structures are *ambiguous* in the sense that there may exist several stable or unstable collective states even under the same external conditions.

An example of a revolutionary situation in a society leading to a bifurcation process can now be given: If two competing totalitarian ideologies, say a and b, are striving to win exclusive influence over the members of the society, the balanced situation of coexistence of a and b is *unstable*. If one of the ideologies has a small initial advantage it will try to stabilize its influence and to increase the number of its followers, for instance by means of psychological and/or economic pressure, until it becomes fully dominant. Hence, there may occur a bifurcation process from a situation of a balanced coexistence of *two* ideologies to a situation in which one ideology is exclusively stabilised. The process of collective opinion formation which could underly the above transition will be described in a simple quantitative model in Chap. 2.

1.3 Significance and Limitations of Quantitative Sociology

This introduction is concluded with a discussion of some epistemological and operative aspects of the relation between quantitative dynamic sociology and the traditional forms of sociology and political science.

On the one hand, the transition from a qualitative to a quantitative formulation of a science has always opened new dimensions of thought and argumentation; this should also apply to the social sciences. On the other hand, the strong structural differences between the natural and social sciences may lead to the (mutually exclusive) objections *either* that quantification definitely is not a successful method for the complex problems of sociology and history *or* that mathematical methods may be successfully applied to these sciences but, in turn, imply the possibility of misuse.

The standpoint of the authors with respect to these views is a cautious one trying neither to overestimate nor to underestimate the role of application of mathematics to sociology. As will be seen in more detail in subsequent chapters, the framework leading (for instance) to an "equation of motion for the socio-configuration" in principle enables a *quantitative* formulation of many originally qualitative concepts: For example, quantifications of attitudes and their flexibility (expressed in terms of transition probabilities and the magnitude of psychological, political and economic activities and trends, which influence the dynamics of a socio-configuration) are possible. The quantification of political trends leads to their *precise definition* but simultaneously to a much more restrictive use of their original qualitative implications. While, for instance, the qualitative concept "adaptability" describes a complex bundle of motivations in individuals, the corresponding parameter describes "adaptability" *only with respect to its average effect* in the equation of motion of the socio-configuration.

The same increase in precision and simultaneous restriction of meaning takes place in the transition from the word "force" of colloquial language to the quantitative concept of "force" in Newtonian mechanics.

A meaningful use of mathematics in the social sciences, however, should take into account the complex structure of social systems from the very beginning. This means that mathematical models of sociological processes, in general, will be too simplified to give a correct *detailed* description of real processes at the microlevel. On the contrary, their *cognitive value* will consist of a *principle insight* into the role and effect of certain trends with respect to the global dynamics of a society and their *operational value* consist of the possibility of using theoretical variants of the models in order to examine hypothetical states in the society. The cognitive and operational value of a model, of course, only exists if the main features of the model actually exist in reality, even if the latter is more complex than the model. Fortunately it can be shown that many models are *structurally stable* with respect to a *large class of perturbations*.

Therefore, in comparing physico-chemical and sociological systems in terms of synergetics the structural and global aspects of dynamic systems are stressed rather than the more detailed aspects of particular models. In this sense it may be better to call the theory of this book *semi-quantitative sociology*.

It will be a decisive test of the intrinsic validity of the concepts described if they can give an explanation not only of the similarities but also of the structural differences between the natural and social sciences: In the natural sciences a closed subdynamics for a given set of a few macrovariables is often found which allows a prediction of their changes in space-time. On the other hand, the predictive power of the traditional social, political and historical sciences is much lower and investigation is often restricted to a *subsequent analysis* and description of historical and political developments.

In terms of the concepts given here the reason for this difference lies in the higher frequency of bifurcations and phase transitions in sociological systems as compared with physico-chemical systems. For instance, the unexpected rise or decline of macrovariables is a frequent process in sociology but a rare event in physics. As a result, in sociology the paths of macrovariables starting from neighbouring initial states may branch out into domains representing completely different situations. In terms of probability distributions an initially sharply peaked distribution will develop into a widely disposed multi-modal situation.

Consequently, the predictability of a future situation starting from a given initial state with a given initial set of macrovariables is, in general, high in physics but much lower in sociology.

From this point of view a conceptual shortcoming of orthodox Marxism becomes evident. A theory of society in terms of the development of classes is a *grossvariable* description according to our definition. It is, therefore, *necessarily probabilistic*. Since processes in society are non-linear, bifurcations and phase transitions in this theory must be expected. These processes are accompanied by critical fluctuations making an unambiguous prediction of the future paths of macrovariables at critical points in time impossible. Such a structure is *incompatible* with the assumption of a strict historical *determinism*! Even the less restrictive assumption that the diverging and bifurcating historical paths of society will asymptotically merge into *one* (the communist) type of society cannot be upheld by general mathematical arguement and is anyway intuitively rather unlikely.

Another important way in which semi-quantitative sociology may allow progress lies in the necessary clear separation between the objective "neutral" description of sociological structures, on the one hand, and the valuation of moral aspects and the estimation of the content of truth of ideologies on the other.

In *qualitative* theories the objective description of structures and the conveyance of convictions are often inseparably connected. In contrast, a *semi-quantitative theory* automatically can only be made functional by describing the structural effects contained in the subdynamics of the quantified macrovariables.

For instance, even mutually exclusive ideologies with – necessarily – opposed intentions and content may lead to comparable effects in terms of certain quantifiable attitudes and material structures in and of a society. In a semi-quantitative model such ideology invariant structures and their dynamics may be identified and verified (or falsified) on an objective basis.

After a clear separation between objective structures and their ethics has been established it is, of course, not only possible but appropriate to make subsequent moral valuations, for instance, to distinguish between desirable or unwanted developments.

In the light of the foregoing discussion one of the main objections that can be raised against an extention of traditional in the direction of a semi-quantitative sociology can be finally considered, namely that the latter cannot remain as the theory of an independent observer but will itself become a constitutive part of the sociological process once perhaps sensationalised model results are widely published. In this way quantitative sociology may distort the attitudes of people and may be abused. This abuse may, for example, consist of a one-sided selection of attitudes or macrovariables and their dynamics for model applications, pretending that this choice is "objective". It could be further argued that a quantitative theory of sociology – because of its definiteness – finally may narrow the scope of individual freedom.

Such arguments certainly must be taken seriously but have to be seen in relation to analogous arguments about other forms of sociological thinking: *All* sociological or historical theories *share the property* not only of analysing but of being part of the sociological process and for that reason being open to positive or negative use in this context. It seems, however, that a *carefully formulated* semi-quantitative theory may have an intrinsic immunity against abuse:

Firstly, there exists the already mentioned automatic separation between the verifiable dynamics of quantitative variables and their moral and emotional evaluation; this may help to restrict the controversy to the non-quantifiable open alternatives and valuations.

Secondly, the choice of macrovariables in a semi-quantitative model is not readily open to manipulation. Only an appropriately chosen set of macrovariables containing the relevant features of the sociological sector considered will lead to a subdynamics in structural agreement with reality.

Thirdly the quantitative dynamic description of certain collective variables in society fortunately does not itself restrict the scope of individuals to make their own decisions; the theory can only *describe* the *collective resultant* of the actions and thoughts of interacting individuals.

2. Opinion Formation – an Elementary Example of Semi-Quantitative Sociology

In this chapter the type of model structure to be developed will be illustrated using an explicit example. The model of opinion formation presented here and first introduced to quantitative sociology by one of the authors [1.36, 37], is an appropriate example for the following reasons:

a) All the relevant types of equation (master equation, Fokker-Planck equation, Langevin equation and mean value equation) and their interrelations can be set up in an elementary, essentially self-contained manner. The illustration therefore serves simultaneously as an introduction to these equations and some of their important properties.

b) The equations can be solved for relevant situations either exactly or (quite well) approximately.

c) The solutions illustrate the important processes of phase transition and bifurcation accompanied by critical fluctuations in a most simple way.

d) The characteristics of the model are stable with respect to detailed and realistic extensions in the modelling of relevant features. It therefore lends itself to a semi-quantitative interpretation of certain political processes.

2.1 The Model

For simplicity within the framework of modelling the dimension of political attitudes, it will be assumed that there exist only two relevant *political opinions*, say 1 and 2, in the model society. These may be represented by two political parties or, on a more comprehensive level, by two competing ideologies.

The problem will be further simplified by assuming a *homogeneous population* whose members have the same individual behaviour probabilities of reactions and interactions in the opinion formation process. The *socio-configuration* $\{n_1, n_2\}$ at a given time t then consists of the numbers n_1 and n_2 of people having opinion 1 and 2, respectively. If the total number of members of the population, say $2N$, is constant the socio-configuration is dependent on only one relevant variable, the integer n, thus:

$$\left.\begin{array}{lll} n_1 + n_2 = 2N & \text{and} \quad n_1 - n_2 = 2n & \text{so} \\ n_1 = N + n & \text{and} \quad n_2 = N - n & \text{where} \\ -N \leq n \leq N & \text{and} \quad 0 \leq n_1, n_2 \leq 2N . & \end{array}\right\} \tag{2.1}$$

The quantities governing the dynamics of opinion formation and containing the essence of the political behaviour of the population are the *individual transition probabilities* per unit time period:

$$\left.\begin{array}{ll} p_{21}(n) & \text{(transition from opinion 1 to 2)} \\ p_{12}(n) & \text{(transition from opinion 2 to 1)} \end{array}\right\} \tag{2.2}$$

The transition probabilities are assumed to be functions of the momentary socio-configuration defined by the variable n. Before considering specific transition probabilities in Sect. 2.4 the mathematical structure of the model in its general form will be discussed in Sect. 2.2 and 2.3, whereby in the following Sect. 2.2 the levels of description adopted for the motion of the system will be explained.

2.2 The Equations of Motion

First an *ensemble* (a statistical population) *of societies* – each of which consists of one homogeneous (human) population of $2N$ members – will be considered and the probability distribution over their socio-configurations will be defined. Thus the function

$$p[n_1, n_2; t] \equiv p(n; t) \tag{2.3}$$

denotes the probability that one sample society has the socio-configuration $\{n_1, n_2\}$ at time t. Of course, the condition

$$\sum_{n=-N}^{N} p(n; t) = 1 \tag{2.4}$$

has to be satisfied at all times.

2.2.1 The Master Equation for $p(n; t)$

The equation of motion for $p(n; t)$ is a special case of the general master equation

$$\frac{dp(i;t)}{dt} = \sum_j [w(i \leftarrow j)p(j;t) - w(j \leftarrow i)p(i;t)] \tag{2.5}$$

already discussed in the Introduction. In this equation $p(i;t)$ is the probability of state i at time t and $w(i \leftarrow j)$ is the transition probability per unit time from state j to state i. The first term of the rhs of (2.5) describes the probability flux from all states j into state i and the second term describes the probability flux from state i into all states j. It can be seen that the generalization of (2.4), viz. the normalization condition

$$\sum_i p(i;t) = 1 \tag{2.6}$$

holds at all times t on the basis of the conservation law:

$$\frac{d}{dt}\sum_i p(i;t) = \sum_{i,j} [w(i \leftarrow j)p(j;t) - w(j \leftarrow i)p(i;t)] \equiv 0 \tag{2.7}$$

which is simply derived from (2.5).

In the case at hand the individual transition probabilities (2.2) induce "nearest neighbour" transitions of the socio-configuration $\{n_1, n_2\}$ only. The transition $\{n_1, n_2\} \to \{n_1 + 1, n_2 - 1\}$, or equivalently $n \to (n+1)$, is effected by a transition from opinion 2 to 1 by *one* of the n_2 population members of opinion 2. Analogously, the transition $\{n_1, n_2\} \to \{n_1 - 1, n_2 + 1\}$, or equivalently $n \to (n-1)$, is effected by a transition from opinion 1 to 2 by *one* of the n_1 population members of opinion 1.

Correspondingly, the transition probabilities for the whole socio-configuration are given by

$$\left.\begin{array}{l} w((n+1) \leftarrow n) \equiv w_\uparrow(n) = n_2 p_{12}(n) = (N-n)p_{12}(n) \\ w((n-1) \leftarrow n) \equiv w_\downarrow(n) = n_1 p_{21}(n) = (N+n)p_{21}(n) \\ w(n' \leftarrow n) = 0 \quad \text{for} \quad n' \neq n \pm 1 \,. \end{array}\right\} \tag{2.8}$$

It is easy to generalize (2.2, 8) to multiple transitions $\{n_1, n_2\} \to \{n_1 \pm m, n_2 \mp m\}$ for the case in which a group of m population members simultaneously changes opinion. The methods of solution of the master equation in this more complicated case, however, will be demonstrated in Chap. 4. According to (2.5, 8) the master equation for the probability is

$$\begin{aligned} \frac{dp(n;t)}{dt} &= [w_\downarrow(n+1)p(n+1;t) - w_\downarrow(n)p(n;t)] \\ &+ [w_\uparrow(n-1)p(n-1;t) - w_\uparrow(n)p(n;t)] \,. \end{aligned} \tag{2.9}$$

An equivalent formulation of (2.9) can be found by introducing probability fluxes:

$$\left.\begin{aligned} j_\uparrow(n;t) &= w_\uparrow(n)p(n;t) \\ j_\downarrow(n;t) &= w_\downarrow(n)p(n;t) \end{aligned}\right\} \qquad (2.10)$$

from "state" n to the states $(n+1)$ and $(n-1)$, where the net flux between the states n and $(n+1)$ is

$$k(n;t) = j_\uparrow(n;t) - j_\downarrow(n+1;t)\ . \qquad (2.11)$$

The master equation then assumes the form of a continuity equation:

$$dp(n;t)/dt = [k(n-1;t) - k(n;t)] \equiv -\Delta k(n;t)\ . \qquad (2.12)$$

From (2.8, 10, 11) the fluxes are subject to the boundary conditions:

$$\left.\begin{aligned} j_\uparrow(N;t) &= j_\downarrow(-N;t) &&= 0 \\ k(N;t) &= k(-N-1;t) &&= 0\ . \end{aligned}\right\} \qquad (2.13)$$

Finally, the master equation (2.5) and, in particular, (2.9) can be written in matrix notation as

$$d|\boldsymbol{p}(t)\rangle/dt = \mathcal{L}|\boldsymbol{p}(t)\rangle \qquad (2.14)$$

where $|\boldsymbol{p}(t)\rangle$ is a vector with components $p(i;t)$, viz.:

$$|\boldsymbol{p}(t)\rangle \equiv \begin{pmatrix} \cdots\cdots \\ \cdots\cdots \\ p(i;t) \\ \cdots\cdots \\ \cdots\cdots \\ \cdots\cdots \end{pmatrix} \qquad (2.15)$$

and $\mathcal{L}$ is a matrix with elements L_{ij}:

$$\mathcal{L} = ((L_{ij})), \qquad \text{where} \quad L_{ij} = w(i \leftarrow j) - \delta_{ij}\sum_k w(k \leftarrow i)\ . \qquad (2.16)$$

In the particular case of (2.9), the vector $|\boldsymbol{p}(t)\rangle$ has $(2N+1)$ components

$$|\boldsymbol{p}(t)\rangle = \begin{pmatrix} p(-N;t) \\ \cdots\cdots \\ p(n;t) \\ \cdots\cdots \\ p(N;t) \end{pmatrix} \qquad (2.17)$$

and the matrix $\mathcal{L}$ can be represented as

$$\left.\begin{aligned} \mathcal{L} &= ((L_{n,n'})) \\ \text{where}\quad L_{n,n} &= -(w_\uparrow(n) + w_\downarrow(n)) \\ L_{n,n-1} &= w_\uparrow(n-1) \\ L_{n,n+1} &= w_\downarrow(n+1) \\ L_{n,n'} &= 0 \qquad \text{for} \quad n' \notin \{n-1, n, n+1\} \,. \end{aligned}\right\} \tag{2.18}$$

2.2.2 The Fokker-Planck Equation for $P(x; t)$

A conversion can now be made from the exact description of the socio-configuration in terms of discrete (integer) numbers n to an approximate description in terms of continuous variables. This will be a good approximation if the integers n_1, n_2, n and N describing the socio-configuration are large and if $p(n; t)$, $w_\uparrow(n)$ and $w_\downarrow(n)$ can be justifiably regarded as continuous functions of n. In the following they will also be considered as differentiable functions of a continuous variable n.

Firstly, the rhs of the master equation (2.9) is expanded as a Taylor series up to and including terms of the second order

$$\begin{aligned} \frac{\partial p(n;t)}{\partial t} = {} & \Delta n \frac{\partial}{\partial n}[w_\downarrow(n)p(n;t)] + \frac{(\Delta n)^2}{2}\frac{\partial^2}{\partial n^2}[w_\downarrow(n)p(n;t)] \\ & - \Delta n \frac{\partial}{\partial n}[w_\uparrow(n)p(n;t)] + \frac{(\Delta n)^2}{2}\frac{\partial^2}{\partial n^2}[w_\uparrow(n)p(n;t)] \end{aligned} \tag{2.19 a}$$

to obtain (setting $\Delta n = 1$):

$$\begin{aligned} \frac{\partial p(n;t)}{\partial t} = {} & -\frac{\partial}{\partial n}\{[w_\uparrow(n) - w_\downarrow(n)]p(n;t)\} \\ & + \frac{1}{2}\frac{\partial^2}{\partial n^2}\{[w_\uparrow(n) + w_\downarrow(n)]p(n;t)\} \,. \end{aligned} \tag{2.19 b}$$

Secondly, the variable x is introduced, where:

$$x = \frac{n}{N}\,; \quad \Delta x = \frac{\Delta n}{N} = \frac{1}{N} \equiv \varepsilon, \qquad \text{so that} \quad -1 \le x \le 1 \,. \tag{2.20}$$

Simultaneously the probability distribution function

$$P(x;t) = Np(n;t) = Np(Nx;t) \tag{2.21}$$

is introduced, which is normalized by

$$\int_{-1}^{+1} P(x;\,t)\,dx \approx \sum_{x=-1}^{+1} P(x;\,t)\,\Delta x = \sum_{n=-N}^{+N} p(n;\,t) = 1\,. \tag{2.22}$$

A "drift coefficient" is defined as:

$$K(x) = W_\uparrow(x) - W_\downarrow(x) \tag{2.23}$$

and a "fluctuation coefficient" as:

$$Q(x) = W_\uparrow(x) + W_\downarrow(x) \tag{2.24}$$

after writing $w_\uparrow(n)$ and $w_\downarrow(n)$ in the form:

$$\left.\begin{aligned} w_\uparrow(n) &= N(1-x)\,p_{12}(Nx) = N\,W_\uparrow(x) \\ w_\downarrow(n) &= N(1+x)\,p_{21}(Nx) = N\,W_\downarrow(x) \end{aligned}\right\}\,. \tag{2.25}$$

Inserting (2.20, 21, 23–25) into (2.19b) the *standard form of a Fokker-Planck equation in one dimension* is obtained:

$$\frac{\partial P(x;\,t)}{\partial t} = -\frac{\partial}{\partial x}\,[K(x)\,P(x;\,t)] + \frac{\varepsilon}{2}\frac{\partial^2}{\partial x^2}\,[Q(x)\,P(x;\,t)] \tag{2.26}$$

which may also be written in the form of a continuity equation

$$\frac{\partial P(x;\,t)}{\partial t} = -\frac{\partial}{\partial x}\,I(x;\,t) \tag{2.27}$$

after introducing the probability flux

$$I(x;\,t) = K(x)\,P(x;\,t) - \frac{\varepsilon}{2}\frac{\partial}{\partial x}\,[Q(x)\,P(x;\,t)]\,. \tag{2.28}$$

Corresponding to (2.13) this probability flux has to satisfy the boundary conditions

$$I(-1;\,t) = I(+1;\,t) = 0\,. \tag{2.29}$$

The meaning of the coefficients $K(x)$ and $Q(x)$ can now be investigated. Starting from an infinitely sharp delta function distribution concentrated at x_0 for time $t = 0$

$$P(x;\,0) = \delta(x - x_0) \tag{2.30}$$

and letting $P(x;\,t)$ develop in the small time interval Δt, the definition of the differential quotient of $P(x; t)$ with respect to time at $t = 0$

$$\frac{\partial P(x;0)}{\partial t} = \lim_{\Delta t \to 0} \frac{P(x;\Delta t) - P(x;0)}{\Delta t} \tag{2.31}$$

will be used. Introducing the notation for the mean value of a function $f(x)$ at time t:

$$\langle f(x) \rangle_t = \int_{-1}^{+1} P(x;t) f(x)\, dx \tag{2.32}$$

and inserting (2.31) in the lhs of (2.26) and taking the first and second moment of (2.26), the short time development of the mean values of x and x^2 can then be calculated.

For the first moment

$$\begin{aligned} \int_{-1}^{+1} \frac{\partial P(x;0)}{\partial t} x\, dx &= \lim_{\Delta t \to 0} \frac{\langle x \rangle_{\Delta t} - \langle x \rangle_0}{\Delta t} \\ &= \int_{-1}^{+1} x \left\{ -\frac{\partial}{\partial x} [K(x) P(x;0)] + \frac{\varepsilon}{2} \frac{\partial^2}{\partial x^2} [Q(x) P(x;0)] \right\} dx \\ &= K(x_0) \end{aligned} \tag{2.33 a}$$

is obtained. The last equation follows after partial integration on the rhs and by using (2.30). Because

$$\langle x \rangle_0 = \int \delta(x - x_0)\, x\, dx = x_0 = \int P(x;\Delta t)\, x_0\, dx = \langle x_0 \rangle_{\Delta t}$$

the result can also be written in the form

$$K(x_0) = \lim_{\Delta t \to 0} \frac{\langle (x - x_0) \rangle_{\Delta t}}{\Delta t}\,. \tag{2.33 b}$$

Similarly it follows from (2.26–32) that

$$\begin{aligned} \int_{-1}^{+1} \frac{\partial P(x;0)}{\partial t} x^2\, dx &= \lim_{\Delta t \to 0} \frac{\langle x^2 \rangle_{\Delta t} - \langle x^2 \rangle_0}{\Delta t} \\ &= \int_{-1}^{+1} x^2 \left\{ -\frac{\partial}{\partial x} [K(x) P(x;0)] + \frac{\varepsilon}{2} \frac{\partial^2}{\partial x^2} [Q(x) P(x;0)] \right\} dx \\ &= 2 x_0 K(x_0) + \varepsilon Q(x_0) \end{aligned} \tag{2.34 a}$$

or, by using (2.33 b) and $x_0^2 = \langle x^2 \rangle_0 = \langle x_0^2 \rangle_{\Delta t}$, that

$$\varepsilon Q(x_0) = \lim_{\Delta t \to 0} \frac{[\langle x^2 \rangle_{\Delta t} - 2x_0 \langle x \rangle_{\Delta t} + x_0^2]}{\Delta t}$$

$$= \lim_{\Delta t \to 0} \frac{\langle (x - x_0)^2 \rangle_{\Delta t}}{\Delta t} . \qquad (2.34\,\text{b})$$

Eqs. (2.33 b, 34 b) show that $K(x_0)$ and $\varepsilon Q(x_0)$ are the mean deviation and mean square deviation of x in the time interval Δt, divided by Δt, which occur if motion is started from the highly peaked initial distribution (2.30). This result will be generalized to the multi-dimensional case in Chap. 3.

2.2.3 The Langevin Equation for $x(t)$

Instead of resorting to the ensemble standpoint the approach can be adopted that *one sample society* is observed and the path of $n(t)$ or of $x(t)$ followed. It is clear that $n(t)$ and $x(t)$ are fluctuating quantities since the increments or decrements in $n(t)$ are integers. For large n, however, the total motion of $n(t)$ and $x(t)$ produced by the stochastic individual transitions adds up to a smooth motion superimposed by small residual fluctuations. This type of motional structure can be described by Langevin equations. Later in Chap. 3 the equivalence between Fokker-Planck and Langevin equations will be derived in general. The essence of the derivation as applied to the present one-dimensional case, however, is the following:

A Langevin equation for the motion of $x(t)$ has the following general form:

$$\frac{dx(t)}{dt} = k(x) + q(x)\,\xi(t) \qquad (2.35)$$

where $k(x)$ and $q(x)$ are to be determined and where $\xi(t)$ are assumed as "Gaussian δ-correlated random forces", i.e. they are assumed to have correlation functions:

$$\langle \xi(t) \rangle = 0 \quad \text{and} \quad \langle \xi(t+\tau)\,\xi(t) \rangle = \delta(\tau) \qquad (2.36)$$

and higher correlation functions which can be expressed uniquely using products of these correlations.

Here the mean value $\langle\ \rangle$ has to be taken over the possible random forces ξ occurring in the sample society via the stochastic transitions of its members. The stochastic nature of the random forces on the rhs of (2.35) in turn leads to non-deterministic, stochastic paths of the variable $x(t)$: Paths $x(t)$ develop differently under the influence of the stochastic forces $\xi(t)$ in (2.35) even if they start out from the same initial value $x(0)$. The mean deviation and mean square deviation of $x(t)$ from the initial value after a short time interval Δt can, however, be calculated by integrating (2.35) iteratively over Δt with the initial condition $x(0) = x_0$ and by afterwards taking the mean value $\langle\ \rangle$. In this way (see Chap. 3 for the explicit procedure) the following is obtained

$$\lim_{\Delta t \to 0} \frac{\langle [x(\Delta t) - x_0] \rangle}{\Delta t} = k(x_0) + \frac{1}{2} q(x_0) \frac{\partial q(x_0)}{\partial x_0} \tag{2.37}$$

and

$$\lim_{\Delta t \to 0} \frac{\langle [x(\Delta t) - x_0]^2 \rangle}{\Delta t} = q^2(x_0) \, . \tag{2.38}$$

This result should be compared with (2.33 b, 34 b) where the short time mean deviation and mean square deviation of x (divided by Δt) are calculated on the basis of the Fokker-Planck equation for $P(x; t)$ and where it turns out that they agree with the drift coefficient and fluctuation coefficient, respectively, of the Fokker-Planck equation. Equating the rhs of (2.33 b, 37) as well as those of (2.34 b, 38), the relations between the coefficients of the Fokker-Planck equation (2.26) and the Langevin equation (2.35)

$$q(x) = \varepsilon^{1/2} Q^{1/2}(x) \tag{2.39}$$

$$k(x) = K(x) - \frac{1}{2} q(x) \frac{\partial q(x)}{\partial x} = K(x) - \frac{\varepsilon}{4} \frac{\partial Q(x)}{\partial x} \tag{2.40}$$

are obtained, which establish the equivalence of the descriptions of societies using the Fokker-Planck equation or the Langevin equation.

2.2.4 Equations for Mean Values

As discussed in the introduction it is quite often sufficient for the solution – except in the case of bifurcation – to establish the paths of the mean values of relevant variables. It is easy to derive equations for $\langle x \rangle_t$ and $\langle x^2 \rangle_t$ from the Fokker-Planck equation (2.26) using the definition (2.32). Multiplying (2.26) with x and x^2 respectively, after partial integration over the interval $[-1,1]$ and taking into account the boundary condition (2.29) the exact equations are obtained as

$$\frac{d\langle x \rangle_t}{dt} = \langle K(x) \rangle_t - \frac{\varepsilon}{2} [Q(x) P(x; t)]_{x=-1}^{x=1} \tag{2.41}$$

and

$$\frac{d\langle x^2 \rangle_t}{dt} = 2 \langle x K(x) \rangle_t + \varepsilon \langle Q(x) \rangle_t - \varepsilon [x Q(x) P(x; t)]_{x=-1}^{x=1} \, . \tag{2.42}$$

It is convenient to introduce the variance

$$\sigma(t) = \langle (x - \langle x \rangle_t)^2 \rangle_t = \langle x^2 \rangle_t - \langle x \rangle_t^2 \geq 0 \,. \tag{2.43}$$

Further, in these mean value equations the boundary contributions can be neglected because very small values of $P(x; t)$ are normally obtained at $x = \pm 1$. Instead of (2.41, 42) the following can then be derived

$$\frac{d\langle x \rangle_t}{dt} = \langle K(x) \rangle_t \tag{2.44}$$

$$\frac{d\sigma(t)}{dt} = 2[\langle x K(x) \rangle_t - \langle x \rangle_t \langle K(x) \rangle_t] + \varepsilon \langle Q(x) \rangle_t \,. \tag{2.45}$$

However, (2.44, 45) for $\langle x \rangle_t$ and $\sigma(t)$ are not closed, except in the simple case when $K(x)$ is linear in x and Q is a constant. In this case (2.26) can also be solved exactly as will be shown in Sect. 2.3.2.

In all other cases *approximate* closed equations for $\langle x \rangle_t$ and $\sigma(t)$ can be used as long as it is justified to assume that the probability distribution $P(x; t)$ has only *one highly peaked maximum*. Under such circumstances $K(x)$ can be expanded in a Taylor series around $\langle x \rangle_t$, as:

$$\begin{aligned} K(x) = {} & K(\langle x \rangle_t) + K'(\langle x \rangle_t)(x - \langle x \rangle_t) \\ & + \frac{1}{2} K''(\langle x \rangle_t)(x - \langle x \rangle_t)^2 + \ldots \end{aligned} \tag{2.46}$$

This expansion can be inserted into the expectation values of the rhs of (2.44, 45), *neglecting higher-order terms*. As $\varepsilon Q(x)$ is proportional to $\varepsilon = 1/N \ll 1$, only the first-order term of $\varepsilon \langle Q(x) \rangle_t$ need be considered. Using:

$$\left.\begin{aligned} \langle x K(x) \rangle_t &= \langle x \rangle_t K(\langle x \rangle_t) + K'(\langle x \rangle_t)\sigma(t) + \ldots \\ \langle K(x) \rangle_t &= K(\langle x \rangle_t) + 1/2\, K''(\langle x \rangle_t)\sigma(t) + \ldots \\ \varepsilon \langle Q(x) \rangle_t &= \varepsilon Q(\langle x \rangle_t) + \ldots \end{aligned}\right\} \tag{2.47}$$

the procedure yields the *closed, but approximate* equations for $\langle x \rangle_t$ and $\sigma(t)$ as:

$$\frac{d\langle x \rangle_t}{dt} = K(\langle x \rangle_t) \tag{2.48}$$

and

$$\frac{d\sigma(t)}{dt} = \varepsilon Q(\langle x \rangle_t) + 2K'(\langle x \rangle_t)\sigma(t) \tag{2.49}$$

In (2.48) the term $1/2\ K''(\langle x\rangle_t)\,\sigma(t)$ has also been neglected since (2.48, 49) will be used for the drift dominated motion only, where $\sigma(t)$ has the order of magnitude of $\varepsilon\, Q(\langle x\rangle_t)$ and where

$$|K(\langle x\rangle_t)| \gg 1/2\ |K''(\langle x\rangle_t)\,\sigma(t)|\ . \tag{2.50}$$

The reader who is not so interested in the derivation of solutions to the equations of motion, but more in the explicit form of the model could skip Sect. 2.3 and continue reading with Sect. 2.4, 5.

2.3 Solutions of the Equations of Motion

Solutions to the master equation, the Fokker-Planck equation and the mean value equations in the stationary and in the time dependent cases will now be derived.

2.3.1 Stationary Solutions

Beginning with the master equation written in the form (2.12) it follows in the stationary case, i.e. for vanishing time derivatives, that:

$$\frac{dp_{st}(n)}{dt} = 0 = [k_{st}(n-1) - k_{st}(n)] \qquad \text{for} \quad -N \le n \le N\ . \tag{2.51}$$

Taking into account the boundary condition (2.13) it can be concluded – by recursive solution of (2.51) starting from $n = -N$ – that

$$k_{st}(n) = 0 \qquad \text{for} \quad -N \le n \le N \tag{2.52}$$

or, by inserting (2.11, 10), that

$$w_\downarrow(n+1)\,p_{st}(n+1) = w_\uparrow(n)\,p_{st}(n) \tag{2.53}$$

where $p_{st}(n)$ is the stationary probability and $k_{st}(n)$ the stationary net flux. Solving (2.53) recursively the *exact stationary solution of the master equation* is obtained as

$$p_{st}(n) = p_{st}(-N) \prod_{\nu=-N+1}^{n} \frac{w_\uparrow(\nu-1)}{w_\downarrow(\nu)} \tag{2.54}$$

for $(-N+1) \le n \le N$

or, equivalently, starting the recursion from $n = 0$, as

$$\left.\begin{aligned} p_{st}(n) &= p_{st}(0)\prod_{\nu=1}^{n}\frac{w_{\uparrow}(\nu-1)}{w_{\downarrow}(\nu)} \quad \text{for} \quad 1 \le n \le N \\ &\text{and} \\ p_{st}(n) &= p_{st}(0)\prod_{\nu=-1}^{n}\frac{w_{\downarrow}(\nu+1)}{w_{\uparrow}(\nu)} \quad \text{for} \quad -N \le n \le -1 \,. \end{aligned}\right\} \tag{2.55}$$

The values of $p_{st}(-N)$ and of $p_{st}(0)$ in (2.54, 55) have to be determined by the normalization condition

$$\sum_{\nu=-N}^{N} p_{st}(\nu) = 1 \,. \tag{2.56}$$

Corresponding to the exact stationary solution (2.54 or 55) of the master equation the *stationary solution $P_{st}(x)$ of the Fokker-Planck equation* can also be determined.

Starting from (2.27) and taking into account the boundary condition (2.29) in the stationary case, viz. when

$$\frac{\partial P_{st}(x)}{\partial t} = 0 = -\frac{\partial}{\partial x} I_{st}(x) \tag{2.57}$$

the result

$$I_{st}(x) = \text{const} = I_{st}(\pm 1) = 0 \tag{2.58}$$

is found, where $I_{st}(x)$ is the stationary probability flux. Inserting (2.28) into (2.58) the first order differential equation for $P_{st}(x)$ is obtained as:

$$\frac{\varepsilon}{2}\frac{\partial}{\partial x}[Q(x)P_{st}(x)] = K(x)P_{st}(x) \tag{2.59}$$

which can be solved immediately. The result is the *exact stationary solution of the Fokker-Planck equation*:

$$P_{st}(x) = P_{st}(x_0)\frac{Q(x_0)}{Q(x)}\exp[N\phi(x)] \tag{2.60}$$

where

$$\begin{aligned} N\phi(x) &= \frac{2}{\varepsilon}\int_{x_0}^{x}\frac{K(x')}{Q(x')}dx' \\ &= \frac{2}{\varepsilon}\int_{x_0}^{x}\frac{W_{\uparrow}(x')-W_{\downarrow}(x')}{W_{\uparrow}(x')+W_{\downarrow}(x')}dx' \quad \text{with} \quad \varepsilon = \frac{1}{N} \,. \end{aligned} \tag{2.61}$$

The normalization condition (2.22) for $P_{st}(x)$ yields

$$P_{st}(x_0) = Q^{-1}(x_0)\left\{\int_{-1}^{+1} Q^{-1}(x')\exp[N\phi(x')]\,dx'\right\}^{-1} . \tag{2.62}$$

The stationary probability density $P_{st}(x)$ has maxima or minima at values x_m defined by

$$\left.\frac{\partial P_{st}(x)}{\partial x}\right|_{x=x_m} = 0 \tag{2.63}$$

and

$$\left.\frac{\partial^2 P_{st}(x)}{\partial x^2}\right|_{x=x_m} \begin{matrix} < 0 & \text{for} \quad x_m = \text{maximum} \\ > 0 & \text{for} \quad x_m = \text{minimum} . \end{matrix} \tag{2.64}$$

If the fluctuation terms proportional to $\varepsilon Q'(x_m)$ and $\varepsilon Q''(x_m)$ in (2.63, 64) are neglected in comparison with the drift terms, which is justifiable for $\varepsilon \ll 1$, (2.63, 64) can be substituted by

$$K(x_m) = 0 \tag{2.65}$$

and

$$\left.\frac{\partial K(x)}{\partial x}\right|_{x=x_m} \begin{matrix} < 0 & \text{for} \quad x_m = \text{maximum} \\ > 0 & \text{for} \quad x_m = \text{minimum} . \end{matrix} \tag{2.66}$$

This implies that apart from small deviations due to the fluctuations the extremal values of the stationary probability $P_{st}(x)$ are found at positions where the "drift force" $K(x)$ vanishes.

2.3.2 *Time Dependent Solutions

1) First the solutions of the equations

$$\frac{dy}{dt} = K(y) = -\frac{\partial V(y)}{\partial y} \tag{2.67}$$

$$\frac{d\sigma}{dt} = \varepsilon Q(y) + 2K'(y)\sigma \tag{2.68}$$

will be considered where $y(t)$ denotes the mean value $\langle x \rangle_t$ and $\sigma(t)$ the variance and where a potential $V(y)$ has been introduced. These equations are valid for sharply peaked uni-modal probability distributions and are, in fact, the same equations as (2.48, 49).

* This section can be omitted on a first reading.

Equation (2.67) also follows from the Langevin equation (2.35) if the fluctuations are neglected by putting $q(x) = 0$, or letting $\varepsilon \to 0$. Therefore (2.67) also describes the *deterministic path*.

Obviously, there exist stationary solutions of (2.67)

$$y(t) = y_m = \text{const} \tag{2.69}$$

at points y_m defined by

$$K(y_m) = 0\,. \tag{2.70}$$

According to (2.65, 66) these points approximately coincide with the maxima/minima of the stationary probability distribution. In the vicinity of y_m, i.e. for $y = y_m + \eta(t)$ with small $\eta(t)$,

$$\frac{d\eta}{dt} = K'(y_m)\,\eta(t) \tag{2.71}$$

is obtained after expanding $K(y)$, whereby a non-vanishing derivative $K'(y_m)$ is assumed, otherwise higher order terms of the expansion of $K(y)$ have to be taken into account. The solutions to (2.71), viz.:

$$\eta(t) = \eta_0 \exp(\gamma t) \qquad \text{where} \quad \gamma = K'(y_m) = -V''(y_m) \tag{2.72}$$

bifurcate from the maxima of the potential, where $\gamma = -V''(y_m) > 0$, and merge into the minima of the potential, where $\gamma = -V''(y_m) < 0$. Therefore the minima (maxima) of the potential $V(y)$ are stable (unstable) stationary points of (2.67).

Between the extrema of $V(y)$, i.e. in domains where $K(y) \neq 0$, the formal solution of (2.67) is

$$\int_{y_i}^{y} \frac{dx}{K(x)} \equiv f(y, y_i) = t\,. \tag{2.73}$$

Solving for y yields $y(t) = f^{(-1)}(y_i; t)$ with an initial value $y(0) = f^{(-1)}(y_i; 0) = y_i$.

In Fig. 2.1 the qualitative form of solutions $y(t)$ to (2.67) is shown for a given potential $V(y)$.

Under the assumption that the solution $y(t)$ to (2.67) has been found from (2.73), it is then easy to find the explicit solution of (2.68), which assumes the form

$$\frac{d\sigma}{dy} = \varepsilon\,\frac{Q(y)}{K(y)} + \frac{2K'(y)}{K(y)}\,\sigma \tag{2.74}$$

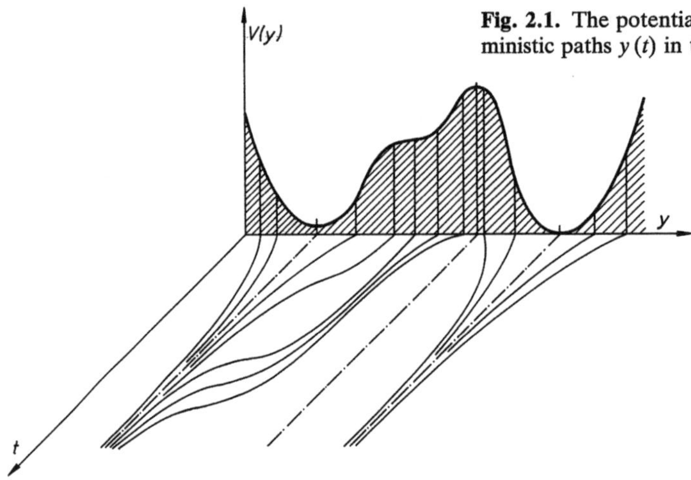

Fig. 2.1. The potential $V(y)$ and several deterministic paths $y(t)$ in the y-t plane

after substituting the variable y for t and $K(y)\frac{d}{dy}$ for $\frac{d}{dt}$. The solution of (2.74) is

$$\begin{aligned}\sigma(t) &= \sigma[y(t)]\\ &= \sigma_0\left[\frac{K(y(t))}{K(y(0))}\right]^2 + K^2(y(t))\int_{y(0)}^{y(t)}\frac{\varepsilon Q(x)}{K^3(x)}\,dx\,.\end{aligned} \tag{2.75}$$

This solution is non-divergent only for $K(y(0)) \neq 0$. Later it will be seen that the solution of (2.68) at an unstable stationary point y_m where $K(y_m) = 0$ and $K'(y_m) = \gamma > 0$, leads to exponential fluctuation enhancement, see (2.93) later in this section.

2) Secondly, the time dependent solutions of the Fokker-Planck equation (2.24) will be studied. Four cases are considered:

a) The case with linear drift and constant fluctuation coefficients for which the exact moment equations (2.44, 45) and the Fokker-Planck equation (2.26) can be solved exactly.

b) The case with drift dominated motion in domains with non-vanishing drift.

c) The case with fluctuation initiated motion starting from a distribution around an unstable stationary point y_m.

d) The case with fluctuation dominated motion starting from a distribution around a stable stationary point y_m.

For each of cases b–d) appropriate approximation methods, which have been introduced by various authors, will be derived. The combination of cases b–d) yields a fairly general picture of the motion of a probability distribution, all the more since the total distribution may be decomposed into a linear superposition of partial distributions. These may be treated *separately* since the Fokker-Planck equation is a *linear equation* for $P(x;t)$.

Case a) **An Exact Solution**. For the following drift and fluctuation coefficients:

$$\left.\begin{aligned} &K(x) = \gamma(x_0 - x) = -\frac{\partial V}{\partial x} \quad \text{for} \quad \gamma > 0 \quad \text{with} \\ &V(x) = \frac{\gamma}{2}(x_0 - x)^2 \quad \text{and} \\ &Q(x) = Q = \text{const} > 0 \end{aligned}\right\} \tag{2.76}$$

the exact moment equations (2.44, 45) assume the forms

$$\frac{d\langle x\rangle_t}{dt} = \gamma(x_0 - \langle x\rangle_t) \tag{2.77}$$

and

$$\frac{d\sigma(t)}{dt} = -2\gamma\sigma(t) + \varepsilon Q \tag{2.78}$$

with solutions

$$\langle x\rangle_t = \langle x\rangle_0\, \mathrm{e}^{-\gamma t} + x_0(1 - \mathrm{e}^{-\gamma t}) \tag{2.79}$$

and

$$\sigma(t) = \sigma(0)\, \mathrm{e}^{-2\gamma t} + \frac{\varepsilon Q}{2\gamma}(1 - \mathrm{e}^{-2\gamma t})\,. \tag{2.80}$$

Further, it can easily be verified that

$$P(x;t) = \frac{1}{\sqrt{2\pi\sigma(t)}} \exp\left[-\frac{(x - \langle x\rangle_t)^2}{2\sigma(t)}\right] \tag{2.81}$$

is the exact and correctly normalized solution to the Fokker-Planck equation (2.26) with the coefficients of (2.76). Equation (2.81) is a normalized Gauss function of width $\sigma^{1/2}(t)$ having its maximum at $x = \langle x\rangle_t$. The mean value and dispersion start from initial values $\langle x\rangle_0$ and $\sigma^{1/2}(0)$ and assume the final values $\langle x\rangle_\infty = x_0$ and $\sigma^{1/2}(\infty) = \sqrt{\varepsilon}\,\sqrt{Q/2\gamma}$.

Case b) **Drift Dominated Motion**. The Fokker-Planck equation in a domain of x, where $K(x) \neq 0$, and for $\varepsilon \ll 1$ will now be considered. The solution (2.75) indicates that in this drift dominated domain the variance $\sigma(t)$ remains of the order of magnitude $O(\varepsilon)$ if the motion is started with an initial variance σ_0 of the same order. This observation suggests that for this domain the variable x can be decomposed into

$$x = y(t) + \varepsilon^{1/2}\xi \qquad \text{with} \quad \xi = \varepsilon^{-1/2}(x - y(t)) \tag{2.82}$$

where $y(t)$ should describe the deterministic path and $\varepsilon^{1/2}\xi$ the deviations from $y(t)$ of order $O(\varepsilon^{1/2})$. It is the aim to derive a simplified equation of motion for

$$\Pi(\xi; t) \equiv P(y(t) + \varepsilon^{1/2}\xi; t) \tag{2.83}$$

in the limit for small ε. The Fokker-Planck equation (2.26) is multiplied by ε and its lhs and rhs expanded in powers of ε. Using (2.82, 83) as well as the notation $\partial/\partial x = \varepsilon^{-1/2}\,\partial/\partial\xi$ the expansions are

$$\text{lhs} = \varepsilon\frac{\partial P(x; t)}{\partial t} = \varepsilon\frac{\partial \Pi(\xi; t)}{\partial t} - \varepsilon^{1/2}\frac{dy}{dt}\frac{\partial \Pi(\xi; t)}{\partial \xi} \tag{2.84}$$

$$\begin{aligned}\text{rhs} = &- \varepsilon^{1/2}\frac{\partial}{\partial\xi}\{[K(y) + K'(y)\,\varepsilon^{1/2}\xi + \dots]\,\Pi(\xi; t)\}\\ &+ \frac{\varepsilon}{2}\frac{\partial^2}{\partial\xi^2}\{[Q(y) + Q'(y)\,\varepsilon^{1/2}\xi + \dots]\,\Pi(\xi; t)\}\,.\end{aligned} \tag{2.85}$$

The comparison of the coefficients of $\varepsilon^{1/2}$ and ε on both sides yields the equations

$$\frac{dy}{dt} = K(y) \tag{2.86}$$

and

$$\frac{\partial \Pi(\xi; t)}{\partial t} = -\frac{\partial}{\partial\xi}[K'(y(t))\,\xi\,\Pi(\xi; t)] + \frac{1}{2}\frac{\partial^2}{\partial\xi^2}[Q(y(t))\,\Pi(\xi; t)]\,. \tag{2.87}$$

The latter equation is a Fokker-Planck equation with linear time dependent drift and time dependent fluctuation. It is easily verified that

$$\begin{aligned}\Pi(\xi; t) &\equiv P(y(t) + \varepsilon^{1/2}\xi; t)\\ &= \frac{1}{\sqrt{2\pi\varepsilon\Sigma(t)}}\exp\left[-\frac{\xi^2}{2\Sigma(t)}\right] = \frac{1}{\sqrt{2\pi\sigma(t)}}\exp\left[-\frac{[x - y(t)]^2}{2\sigma(t)}\right]\end{aligned} \tag{2.88}$$

is a solution to (2.87) if $\Sigma(t) = \varepsilon^{-1}\sigma(t)$ satisfies:

$$\frac{d\Sigma}{dt} = 2K'(y(t))\,\Sigma + Q(y(t)) \tag{2.89}$$

This remarkable result, derived by *van Kampen* [2.1] and others [1.33] implies:

a) that in the drift dominated domain and in the limit for small ε the equations (2.67, 86) and (2.68, 89) for the mean value and variance become exact;

b) that, vice versa, in this limit the mean value $y(t)$ and the variance $\sigma(t)$ are sufficient to determine the equation of motion (2.87) and its solution (2.88) for the probability distribution.

Case c) **Fluctuation Initiated Motion**. If motion is started from a distribution concentrated around an unstable stationary point x_0 where $K(x_0) = 0$ and $K'(x_0) = \gamma > 0$, the expansion of (2.82, 84, 85) fails since the initial fluctuations are enhanced exponentially, see (2.92) before the drift dominated motion sets in. The appropriate approximation here, developed by *Haake* [2.2, 3] and by *Suzuki* [2.4] essentially consists of two steps: 1) Solve the Fokker-Planck equation for the first fluctuation dominated stage and 2) Find a smoothly fitting solution for the second drift dominated stage.

In the vicinity of the unstable point x_0 the Fokker-Planck equation (2.26) may be approximated by

$$\frac{\partial P(x;t)}{\partial t} = -\frac{\partial}{\partial x}[K'(x_0)(x-x_0)P] + \frac{1}{2}\frac{\partial^2}{\partial x^2}[\varepsilon Q(x_0)P] \tag{2.90}$$

with the solution

$$P(x;t) = \frac{1}{\sqrt{2\pi\varepsilon\Sigma(t)}} \exp\left[-\frac{(x-x_0)^2}{2\varepsilon\Sigma(t)}\right] \tag{2.91}$$

where the variance $\sigma(t) = \varepsilon\Sigma(t)$ is such that

$$\sigma(t) = \varepsilon\left(\Sigma(0) + \frac{Q(x_0)}{2\gamma}\right)e^{2\gamma t} - \varepsilon\frac{Q(x_0)}{2\gamma} \tag{2.92}$$

which, with $\gamma = K'(x_0) > 0$ is the solution of the equation

$$\frac{d\sigma(t)}{dt} = 2K'(x_0)\sigma(t) + \varepsilon Q(x_0)\,. \tag{2.93}$$

Of course, (2.93) is the variance equation (2.68) at the unstable point $x = x_0$. It is observed that the variance (2.92) of the distribution increases exponentially, starting with an order of magnitude $O(\varepsilon)$ and reaching magnitude $O(1)$ within a fluctuation enhancement period. After this the drift-dominated period sets in, for which the full drift term is needed but from which point on the fluctuation term in the Fokker-Planck equation can be neglected by writing

$$\frac{\partial}{\partial t}P(x;t) + \frac{\partial}{\partial x}[K(x)P(x;t)] = 0 \tag{2.94}$$

or, with $P(x;\, t) = P[x;\, \tau]$

$$2\gamma\tau\frac{\partial}{\partial\tau} P[x;\,\tau] + K(x)\frac{\partial}{\partial x} P[x;\,\tau] = -K'(x)\,P[x;\,\tau] \tag{2.95}$$

after introducing the new time variable

$$\tau = \varepsilon\left(\Sigma(0) + \frac{Q(x_0)}{2\gamma}\right) e^{2\gamma t} \tag{2.96}$$

which is suggested by the form of the solution (2.91) for the first stage. It remains to find that solution of (2.95) for the second stage which smoothly fits onto the solution (2.91) for the first stage.

The characteristic equations belonging to the quasi-linear partial differential equation (2.95) are

$$\left.\begin{aligned} \frac{d\tau}{ds} &= 2\gamma\tau \\ \frac{dx}{ds} &= K(x) \\ \frac{dP}{ds} &= -K'(x)\,P\,. \end{aligned}\right\} \tag{2.97}$$

The elimination of s leads to

$$\left.\begin{aligned} \frac{dx}{K(x)} &= \frac{d\tau}{2\gamma\tau} \\ &\text{and} \\ \frac{dP}{P} &= -\frac{K'(x)}{K(x)}\,dx \end{aligned}\right\} \tag{2.98}$$

with solutions

$$\tau = \tau_0 \exp\left[2\gamma\int_a^x \frac{dx'}{K(x')}\right] \equiv \tau_0\,\varphi^2(x) \tag{2.99}$$

and

$$P(x,\,\tau_0) = \frac{1}{K(x)}\,G(\tau_0) \tag{2.100}$$

where $G(\tau_0)$ is an arbitrary function of the integration constant τ_0 and the constant a can be chosen so that

$$\varphi(x) = (x - x_0) + \dots \quad \text{and} \quad \varphi'(x) = 1 + \dots$$

The solution $P[x; \tau]$ to (2.95) arises from (2.100) after expressing τ_0 as a function of x and τ:

$$P[x; \tau] = P(x; \tau_0(x, \tau)) = \frac{1}{K(x)} G\left(\frac{\tau}{\varphi^2(x)}\right). \tag{2.101}$$

It is clear that a choice of

$$G(\tau_0) = \frac{\gamma}{\sqrt{2\pi\tau_0}} \exp\left(-\frac{1}{\tau_0}\right) \tag{2.102}$$

leads to a solution

$$P[x; \tau] = \frac{1}{\sqrt{2\pi\tau}} \varphi'(x) \exp\left[-\frac{\varphi^2(x)}{\tau}\right] \tag{2.103}$$

for the second stage which develops smoothly out of the fluctuation initiated first stage solution (2.91). It can be shown by considering explicit examples that (2.103) describes the development from the uni-modal distribution at the unstable point to a bimodal distribution running down the slopes of the potential $V(x)$. However, it is worth mentioning that (2.103) does not describe the effect of fluctuations in the final stage of the path when the distribution has reached the neighbouring minima of the potential $V(x)$ at x_+, x_- where $K(x_\pm) = 0$ and $K'(x_\pm) < 0$. This is because fluctuations in the second stage have been neglected by using (2.94) instead of (2.26) which is easily seen from the conservation law

$$\begin{aligned} \frac{\partial}{\partial t} \int_{x_-}^{x_+} P(x; t)\, dx &= -\int_{x_-}^{x_+} \frac{\partial}{\partial x} [K(x) P(x; t)]\, dx \\ &= -[K(x) P(x; t)]_{x_-}^{x_+} = 0\,. \end{aligned} \tag{2.104}$$

From (2.104) it follows that a probability distribution obeying (2.94) and initially confined between x_- and x_+ will never overlap x_+ or x_-.

Case d) **Fluctuation Dominated Motion**. An initial distribution $P(x; 0)$ concentrated at the potential minima x_- and x_+ neighbouring the unstable point x_0 will now be considered. In this case a slow equilibration process will take place by means of a probability flux between the modes until the stationary distribution $P_{st}(x)$ is established. Since the drift force $K(x)$ is directed towards the minima x_- and x_+, it cannot be primarily responsible for this process. In fact the motion

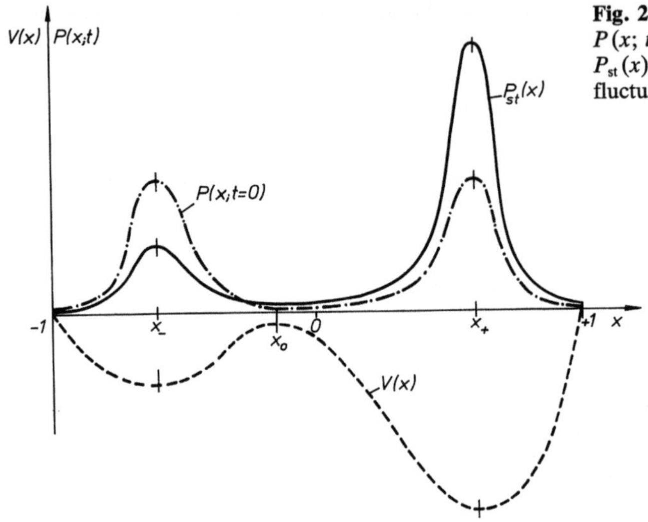

Fig. 2.2. The initial distribution $P(x; t = 0)$, the final distribution $P_{st}(x)$ and the potential $V(x)$ for fluctuation dominated motion

is fluctuation dominated and has been treated by *Kramers* and others [2.5, 6] (see also [1.14]). The equilibration process is shown in Fig. 2.2.

An equation of motion for the probabilities

$$\pi_+(t) = \int_{x_0}^{1} P(x; t)\, dx \qquad \text{and} \quad \pi_-(t) = \int_{-1}^{x_0} P(x; t)\, dx \tag{2.105}$$

will now be established. Using the Fokker-Planck equation in the form (2.27) together with the boundary condition (2.29)

$$\left.\begin{aligned} \frac{d\pi_+}{dt} &= -\int_{x_0}^{1} \frac{\partial}{\partial x} I(x; t)\, dx = I(x_0; t) \\ \text{and} & \\ \frac{d\pi_-}{dt} &= -\int_{-1}^{x_0} \frac{\partial}{\partial x} I(x; t)\, dx = -I(x_0; t) \end{aligned}\right\} \tag{2.106}$$

are obtained. Before evaluating these equations a convenient form for the probability current $I(x; t)$ will be stated

$$\frac{2I(x; t)}{Q(x)\, P_{st}(x)} = -\varepsilon \frac{\partial}{\partial x}\left(\frac{P(x; t)}{P_{st}(x)}\right). \tag{2.107}$$

The equivalence of this with the original definition (2.28) for $I(x; t)$ can be easily proved by inserting the explicit stationary solution (2.60) for $P_{st}(x)$ in (2.107). Furthermore the highly plausible assumption will be now made that the probability current between the probability modes at x_- and x_+ is independent of the position x during the equilibration process

This means that

$$I(x;\, t) = I(t) \qquad \text{for} \quad x_- \leq x \leq x_+ \, . \tag{2.108}$$

Integrating (2.107) from x_- to x_+ under the assumption (2.108) yields:

$$I(t) \;=\; -\gamma\left(\frac{P(x_+;\, t)}{P_{st}(x_+)} - \frac{P(x_-;\, t)}{P_{st}(x_-)}\right) \tag{2.109}$$

with

$$\gamma \;=\frac{\varepsilon}{2}\left\{\int\limits_{x_-}^{x_+} [Q(x)\, P_{st}(x)]^{-1}\, dx\right\}^{-1} \tag{2.110}$$

Since the probability is concentrated around x_+ and x_- for the whole time evolution it seems also reasonable to assume:

$$\frac{P(x_\pm;\, t)}{P_{st}(x_\pm)} = \frac{\pi_\pm(t)}{\pi_{\pm st}} \, . \tag{2.111}$$

Inserting (2.109, 111) into the rhs of (2.106) it can finally be seen that

$$\frac{d\pi_+}{dt} \;=\; -\frac{d\pi_-}{dt} \;=\; -\gamma\left(\frac{\pi_+}{\pi_{+st}} - \frac{\pi_-}{\pi_{-st}}\right) \tag{2.112}$$

or

$$\left.\begin{aligned} &\frac{d\pi}{dt} = 2\Gamma[\pi_{st} - \pi(t)] \\ \text{with} \quad &\pi(t) \;=\; \pi_+(t) - \pi_-(t) \\ \text{and} \quad &\Gamma \;=\; \frac{\gamma}{2\pi_{+st}\,\pi_{-st}} \, . \end{aligned}\right\} \tag{2.113}$$

In order to calculate the rate Γ or transition time $\tau = \Gamma^{-1}$ the stationary solution (2.60) has to be inserted into (2.110). Considering the symmetrical case where the relations

$$\left.\begin{aligned} &K(-x) \;=\; -K(x) \\ &Q(-x) \;=\; Q(x) \\ &P_{st}(-x) \;=\; P_{st}(x) \\ &\pi_{+st} \;=\; \pi_{-st} \;=\; 1/2 \\ &x_0 \;=\; 0, \qquad x \;=\; x_m \;=\; -x_- \end{aligned}\right\} \tag{2.114}$$

hold good, the transition time is obtained as:

$$\tau = \Gamma^{-1} = N\left\{\int_{-x_m}^{+x_m} \exp[-N\phi(x')]\,dx'\right\}\left\{\int_{-1}^{+1} Q^{-1}(x'')\exp[N\phi(x'')]\,dx''\right\}. \tag{2.115}$$

Evaluating the integrals by the method of steepest descend, the solution

$$\tau = \frac{4\pi Q^{-1}(x_m)}{\sqrt{\phi''(0)|\phi''(x_m)|}}\exp[N\phi(x_m)] \tag{2.116}$$

is obtained. Evidently the transition time τ for the distribution to move from x_- to x_+ is extremely long because of the exponential dependence on $N \gg 1$. The time τ has to be compared with the time in which the deterministic path traverses a distance Δx in the drift dominated domain. According to (2.59) the latter is given by:

$$\Delta t = \int_{y_1}^{y_1+\Delta x} \frac{dx'}{K(x')}. \tag{2.117}$$

2.4 Choice of Transition Probabilities and the Explicit Form of the Model

In Sect. 2.2, 2.3 various versions of the equations of motion and their solutions for the general model have been considered *without reference* to a definitive choice of the decisive transition probabilities $p_{ij}(n)$ defined by (2.2).

In order to obtain *explicit results* an *explicit form* for the $p_{ij}(n)$ has to be chosen. In principle two procedures appear to be possible:

a) $p_{ij}(n)$ could be determined *experimentally* for an ensemble of people under certain opinion formation conditions.

b) A plausible *theoretical* form for $p_{ij}(n)$ could be developed. This form should be flexible enough to describe a manifold of possibilities; i.e. it must contain one or more open parameters. Inserting the form developed into the equations of the model results in certain structures for the socio-configuration. These could then be compared with reality in order to verify or negate the plausibility of the chosen transition probabilities.

The first alternative is the more direct but perhaps also the more difficult method: The process of political opinion formation is a complex one and it is difficult to extract the dependence of opinion changes on the existing opinion distribution under realistic conditions in an experimentally accessible form. The second alternative will therefore be used here.

The transition probabilities to be developed for the individual members of a society in a collective opinion formation process should at least make some use of the following general psychologically and politically orientated observations:

a) There exist individual *preferences* for one of the two opinions assumed in the model which are *independent* of the opinions of other people. Such preferences may be due to inherent inclination and personal tradition or personal observation and estimation of a situation.

b) On the other hand there also exists a certain willingness for the individual to adapt to the prevailing opinion once the latter becomes more and more preponderant. Such an *adaptation trend* may be due to an inherent adaptiveness of the individual or to external pressure influencing the individual readiness to decide in favour of the opinion of the majority.

c) Preference and adaptation trends are superimposed in each individual. Both trends could have independent and different strengths in a given situation.

As one homogeneous population only is to be considered in the simple model presented here, the same strengths for trends for each individual member of the population are assumed. The following forms [1.36, 37] for $p_{ij}(n)$ take into account the observations a–c):

$$\left.\begin{aligned} p_{12}(n) &= \nu \exp(\delta + \tilde{\kappa} n) = \nu \exp(\delta + \kappa x) \\ p_{21}(n) &= \nu \exp[-(\delta + \tilde{\kappa} n)] = \nu \exp[-(\delta + \kappa x)] \end{aligned}\right\} \tag{2.118}$$

with $\kappa = N\tilde{\kappa}$ where δ, κ and ν are *trend parameters* whose effect on the transition probabilities can be described as follows:

a) Clearly, a positive δ increases the probability that an individual changes form opinion 2 to opinion 1 and reduces the probability of changing from 1 to 2 and vice versa for negative δ. δ is therefore denoted as the *preference parameter*.

b) Remembering the definition of n and x it is clear that a positive κ enlarges the transition probability in favour of the majority opinion and reduces the transition probability in the direction of the minority opinion. This effect grows for growing $|x|$ or for a growing imbalance of opinions. Through the term κx the probabilities $p_{ij}(n)$ depend on the socio-configuration. Because of its effect κ is denoted as the *adaptation parameter*.

c) The preference and adaptation trends are automatically superimposed in (2.118). Furthermore, the transition probabilities are positive definite by definition.

d) The parameter ν determines the frequency of the occurring flipping processes or, in other words, the time scale in which opinion changes occur. ν is denoted as the *flexibility parameter*.

Having reached a decision for the transition probabilities it is now possible to make the model explicit by inserting (2.118) into all the equations. In the master equation (2.9) for the socioprobability $p(n; t)$, $w_{\uparrow}(n)$ and $w_{\downarrow}(n)$ have to be used in the form given by:

$$\left.\begin{aligned} w_\uparrow(n) &= \nu(N-n)\exp(\delta + \tilde{\kappa}n) \\ w_\downarrow(n) &= \nu(N+n)\exp[-(\delta + \tilde{\kappa}n)] \ . \end{aligned}\right\} \tag{2.119}$$

For the Fokker-Planck equation (2.26) and using

$$\left.\begin{aligned} W_\uparrow(x) &= \nu(1-x)\exp(\delta + \kappa x) \\ W_\downarrow(x) &= \nu(1+x)\exp[-(\delta + \kappa x)] \end{aligned}\right\} \tag{2.120}$$

the drift coefficient

$$\begin{aligned} K(x) &= -\frac{\partial V(x)}{\partial x} = W_\uparrow(x) - W_\downarrow(x) \\ &= 2\nu[\sinh(\delta + \kappa x) - x\cosh(\delta + \kappa x)] \end{aligned} \tag{2.121}$$

belonging to the potential

$$V(x) = \frac{2\nu}{\kappa^2}[\kappa x \sinh(\delta + \kappa x) - (1+\kappa)\cosh(\delta + \kappa x)] + \text{const} \tag{2.122}$$

and the fluctuation coefficient

$$\begin{aligned} Q(x) &= W_\uparrow(x) + W_\downarrow(x) \\ &= 2\nu[\cosh(\delta + \kappa x) - x\sinh(\delta + \kappa x)] \end{aligned} \tag{2.123}$$

are obtained.

These coefficients can now be inserted into (2.48, 49) for the mean value $\langle x\rangle_t$ and the variance $\sigma(t)$. This leads to the coupled equations

$$\begin{aligned} \frac{d\langle x\rangle}{dt'} &= 2[\sinh(\delta + \kappa\langle x\rangle) - \langle x\rangle\cosh(\delta + \kappa\langle x\rangle)] \\ \frac{d\Sigma(t')}{dt'} &= 2[\cosh(\delta + \kappa\langle x\rangle) - \langle x\rangle\sinh(\delta + \kappa\langle x\rangle)] \\ &\quad -4\Sigma[(1-\kappa)\cosh(\delta + \kappa\langle x\rangle) + \kappa\langle x\rangle\sinh(\delta + \kappa\langle x\rangle)] \end{aligned} \tag{2.124}$$

$$\text{where} \quad t' = \nu t \quad \text{and} \quad \Sigma(t') = N\sigma(t') \tag{2.125}$$

which remain valid as long as the distribution stays uni-modal (Sect. 2.3). The scaled time variable $t' = \nu t$ has been introduced for which the prime will now be omitted.

If (2.119) is used in (2.55) the product expressions can be evaluated and the explicit exact stationary solution to the master equation obtained as:

$$p_{\mathrm{st}}(n) = p_{\mathrm{st}}(0)\,\frac{(N!)^2}{(2N)!}\binom{2N}{N+n}\exp\left[2\delta n + \tilde{\kappa} n^2\right]. \tag{2.126}$$

A further evaluation of the binomial coefficients and the other factorials using Stirling's formula

$$\ln(M!) \approx M \ln M - M \tag{2.127}$$

yields (with $n = Nx$) a more convenient form for $p_{\mathrm{st}}(n)$ as

$$p_{\mathrm{st}}(Nx) = p_{\mathrm{st}}(0)\exp\left[N\,U(x)\right] \tag{2.128}$$

where

$$U(x) = 2\delta x + \kappa x^2 - \left[(1+x)\ln(1+x) + (1-x)\ln(1-x)\right]. \tag{2.129}$$

The extrema of this distribution are determined by

$$\left.\frac{\partial U(x)}{\partial x}\right|_{x=x_m} = 2\delta + 2\kappa x_m - \left[\ln(1+x_m) - \ln(1-x_m)\right] = 0$$

or

$$x_m = \tanh(\delta + \kappa x_m). \tag{2.130}$$

The stationary solution of the Fokker-Planck equation gives practically equivalent results. Inserting into (2.60):

$$P_{\mathrm{st}}(x) = N p_{\mathrm{st}}(Nx) = P_{\mathrm{st}}(x_0)\,\frac{Q(x_0)}{Q(x)}\exp\left[N\phi(x)\right]$$

the fluctuation coefficient $Q(x)$ given by (2.123) as well as

$$\phi(x) = 2\int_{x_0}^{x}\frac{\sinh(\delta + \kappa x') - x'\cosh(\delta + \kappa x')}{\cosh(\delta + \kappa x') - x'\sinh(\delta + \kappa x')}\,dx' \tag{2.131}$$

it can be seen that for large N the extrema of $P_{\mathrm{st}}(x)$ are determined by:

$$K(x_m) = 2\nu\left[\sinh(\delta + \kappa x_m) - x_m\cosh(\delta + \kappa x_m)\right] = 0 \tag{2.132}$$

and this equation agrees with (2.130).

The graphical solution of the transcendental equation (2.130) leads to only one possible solution x_m for arbitrary δ, when $0 < \kappa < 1$. This solution corresponds to the *maximum* of the stationary distribution $P_{\mathrm{st}}(x)$. On the other hand there still exists one solution x_m when $\kappa > 1$ and $|\delta| > \delta_0$, but three solutions $x_- < x_0 < x_+$ when $\kappa > 1$ and $|\delta| < \delta_0$, where δ_0 is determined by the equation

$$\cosh^2[\delta_0 - \sqrt{\kappa(\kappa-1)}] = \kappa \tag{2.133}$$

For $|\delta| = \delta_0$, two of the three solutions x_+, x_0, x_- coincide at

$$x_c = \pm\sqrt{\frac{\kappa-1}{\kappa}}\,. \tag{2.134}$$

In the case with three solutions *two maxima* of $P_{st}(x)$ are found at x_+ and x_- and *one minimum* at x_0.

In the Figs. 2.3 a–d the graphical solution to (2.130) is plotted for several typical choices of κ and δ.

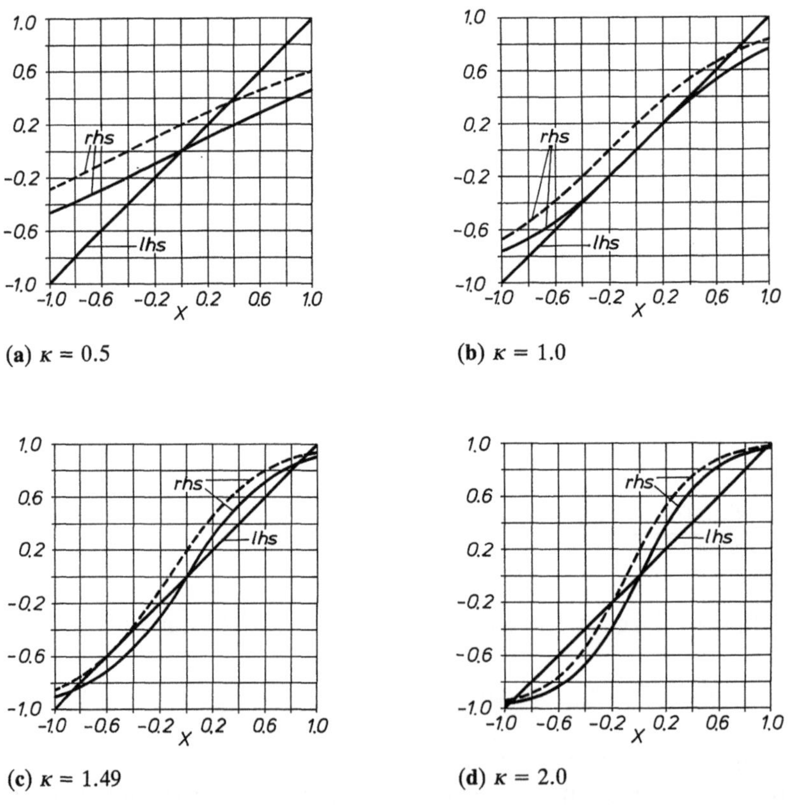

(a) $\kappa = 0.5$ (b) $\kappa = 1.0$

(c) $\kappa = 1.49$ (d) $\kappa = 2.0$

Fig. 2.3 a–d. Graphical solutions of the transcendental equation $x = \tanh(\delta + \kappa x)$ in typical cases. (—) $\delta = 0$; (– – –) $\delta = 0.2$

2.5 The Sociological Interpretation of the Model

Some semi-quantitative conclusions about political opinion formation in societies [1.36, 47] can now be made.

Despite the apparent oversimplications in the model, the main assumptions of only two opinions and of one homogeneous population described by the same trend parameters seem to find some justification: In liberal societies a trend to *two main parties*, usually representing "progressive" and "conservative" opinions, can often be observed. In totalitarian societies a struggle between *two main ideologies*, usually the ruling one trying to defend its "achievements and convictions" and a "counter revolutionary" one trying to win influence, can be seen. Further, modern media tend to give a certain homogeneity to the information available in a society of a given type and this leads to a homogeneity of opinion formation trends – at least for a large proportion of the population.

A general problem consists in the fact that the master equation and the Fokker-Planck equation describe the motion and shape of *probability distributions over an ensemble of societies*. Of course, only *one sample*, namely *the existing society*, is under consideration whose fate corresponds to *one Langevin path*. It is known, however, that the mode (or modes) of a distribution $P(x; t)$ correspond to the most probable position (positions) of the Langevin path $x(t)$. Further, the dispersion of the distribution pertains to the possible random fluctuations and deviations of the Langevin path $x(t)$ from the mean path $\langle x \rangle_t$. With regard to the stationary ensemble it has been seen that the frequency of finding the Langevin path in a space element dx is proportional to the probability $P_{st}(x)\,dx$. Two main cases will now be considered:

a) It is assumed that the trend parameters δ, κ and ν remain constant for a long time in comparison with the evolution time of the political opinion configuration. Under this condition equilibrium distributions of the socio-configuration pertaining to appropriate sets of parameters δ, κ, ν will be discussed together with the development of non-equilibrium socio-configurations for constant δ, κ and ν.

b) The trend parameters δ, κ, ν describing the individual trends of opinion formation may themselves become functions of time. Some plausible cases and their consequences will also be considered under such conditions.

Case a). An interpretation is based on the Figures 2.4–6 which show stationary distributions $p_{st}(n)$ according to (2.128, 129) for chosen values of κ and δ (ν does not appear in $p_{st}(n)$). The relatively small number $N = 25$ has been chosen in order to obtain a conveniently large dispersion of the distributions for illustrative purposes. The choice of a very large N would not change $U(x)$, but the dispersion of the distributions around their modes would shrink proportional to $N^{-1/2}$.

In the case when $\kappa = 0.5$ shown in Fig. 2.4 an interpretation in terms of a *liberal society* is possible because the adaptation trend to the prevailing opinion is low. The independent preferences of the individuals described by δ play the main role. For $\delta = 0$, i.e. no preference, the most probable society (the mode of

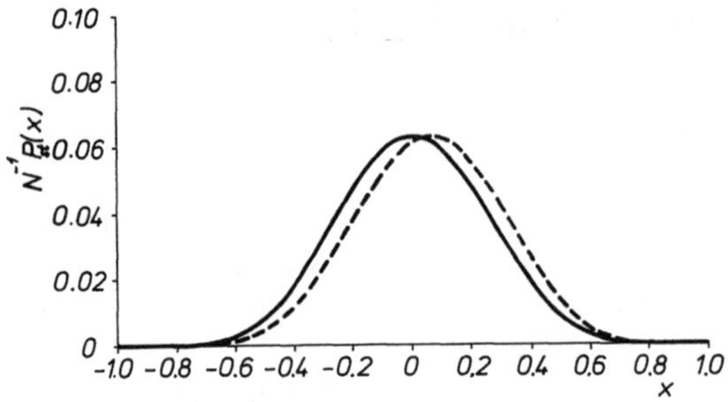

Fig. 2.4. Distribution $P_{st}(x)$ for $N = 25$, $\kappa = 0.5$, $\delta = 0$ (—) and $\kappa = 0.5$, $\delta = 0.02$ (---)

the probability distribution) is one with $n = 0$ or $n_1 = n_2$, i.e. with an equal frequency of the opinions 1 and 2. The dispersion of the distribution is a measure of the fluctuations around this mean state. When $\delta = 0.02$, i.e. in the case of individual preferences for opinion 1, the mean value is shifted in this direction while the shape of the distribution remains approximately the same.

When $\kappa = 1$ or $\kappa = 1.07$ as shown in Fig. 2.5 an interpretation in terms of a society at the crossroads between a *liberal* and a *totalitarian state* can be made. The reason for this is the higher willingness of the individual to adapt to the prevailing opinion. As can be seen in Figs. 2.3 b and c and Fig. 2.5 the value $\kappa_c = 1$ is the *critical value* of the trend parameter κ, where for $\delta = 0$ the potential (2.129) changes from the monostable to the bistable form when all the possible consequences for bifurcation etc., as discussed in Sect. 2.3, can be observed.

When $\delta = 0$ there still occurs a distribution with a maximum around $x = 0$, i.e. the most probable society still has a balanced opinion distribution. However, the dispersion of the distribution is now very large. This implies that large and persistent fluctuating deviations from the balanced situation can be expected. The reason for this is *not* the individual preference, since $\delta = 0$, but the readiness to align to the momentarily prevalent opinion.

In the case when $\kappa = 1.07$ and $\delta = 0.02$ it is observed that the mean value of the distribution in Fig. 2.5 is shifted further towards the boundary $x = +1$ than in Fig. 2.4, although the same preference parameter $\delta = 0.02$ has been used. This means that the opinion alignment effect shifts the mean value of the socio-configuration further into the prevailing direction than without mutual adaptation.

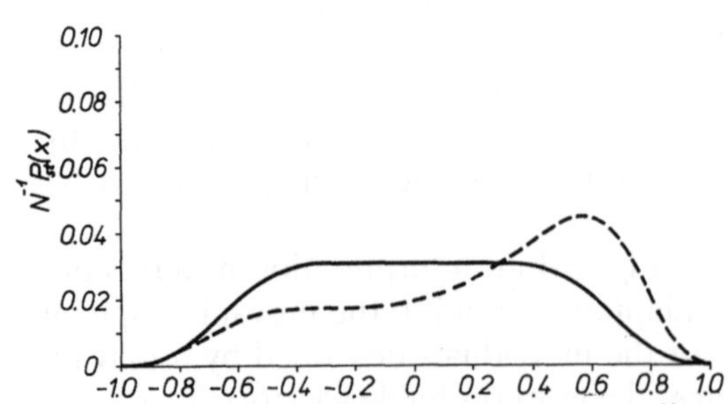

Fig. 2.5. Distribution $P_{st}(x)$ for $N = 25$, $\kappa = 1.0$, $\delta = 0$ (—) and $\kappa = \kappa_0 = 1.07$, $\delta = 0.02$ (---)

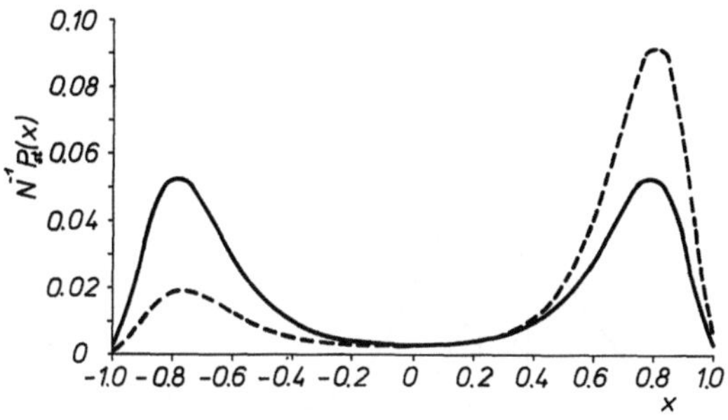

Fig. 2.6. Distributions $P_{st}(x)$ for $N = 25$, $\kappa = 1.3$, $\delta = 0$ (—) and $\kappa = 1.3$, $\delta = 0.02$ (---)

The case when $\kappa = 1.3$ (Fig. 2.6) can be interpreted in terms of a society in a *totalitarian state* with high adaptation willingness (whatever the motivations may be) to the prevailing opinion. Some perhaps remarkable consequences can be deduced: Even in the case $\delta = 0$ (with no individual preferences) the society with a *balanced opinion configuration* is an *improbable one*. In fact societies with strongly prevailing opinions 1 or 2 and with an *asymmetric opinion configuration* around x_+ or x_- are the *most probable*. Evidently this effect is due to the collective self-stabilisation of one opinion through the mutual adaptation of the members of the society. Clearly, this effect is independent of the "truth" content of the opinions 1 or 2.

In the case when $\delta = 0.02$ the expected large probability peak at x_+ corresponding to the prevalence $\delta > 0$ is found. However, its maximum is shifted more towards the boundary $x = +1$ than for $\kappa = 0.5$ as can be seen by comparison with Fig. 2.4; the strong adaptation trends present exaggerate a given preference and shift the opinion configuration to more extreme values.

Even more interesting is the small probability peak at $x_- < 0$. It could be assumed that – because of, say, previous history – the sample society does not belong to the large but to the small probability peak. (Later the consideration of the temporal change of trend parameters shows that this is indeed a relevant case.) This would imply that in this society a collectively stabilized opinion configuration $x_- < 0$ exists in contradiction to the sign and magnitude of the individual preference parameter $\delta > 0$ of all its individual members. A naive expectation could be that such a society would quickly move to the domain of the large probability peak. However, (2.110) evaluated for the present case shows that this event is *extremely* unlikely. In other words, as long as the trend parameters, in particular the high adaptation parameter $\kappa > 1$, remain constant and no external effects disturb or perturb the society, a collective opinion configuration in the totalitarian state of a society persists by self-sustainment for a long time even if individual preferences are in disagreement with the collectively stabilized opinion. In order to give an account of such an evolution Figs. 2.7 and 8 show a solution of the model in terms of the discrete master equation, where $N = 25$ has again been chosen for the illustration. The initial distributions ($t \leq 0$) in Figs. 2.7 and 2.8 are assumed to be equilibrium distributions corresponding to the system parameters $\kappa_0 = 0.5$, $\delta_0 = 0$ and $\kappa_0 = 0.5$, $\delta_0 = -0.15$ respectively. At time $t = 0$ a sudden change of the parameters to the values $\kappa = 1.5$ and $\delta = 0$ is made. Figs. 2.7 and 2.8 specifically show the relaxation of

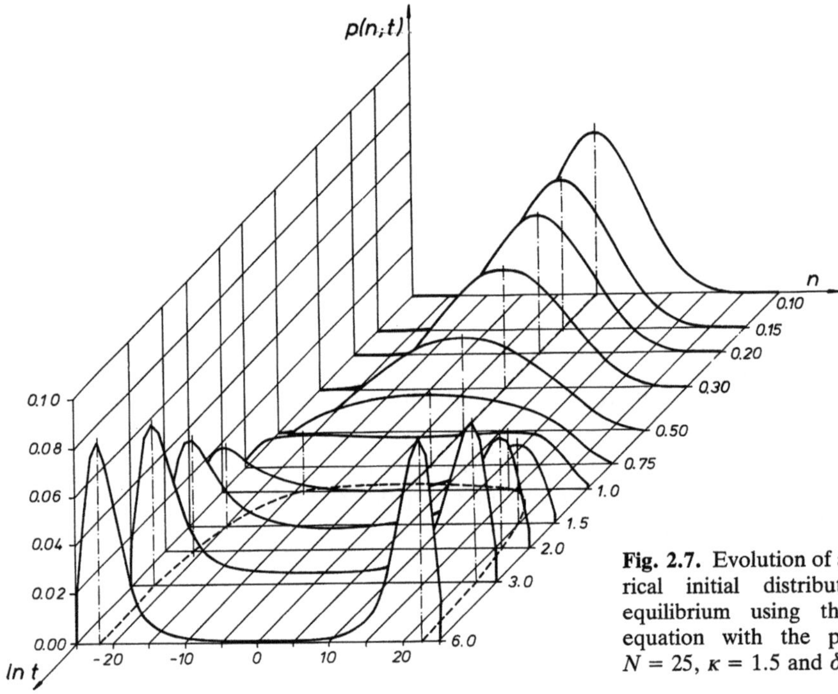

Fig. 2.7. Evolution of a symmetrical initial distribution into equilibrium using the master equation with the parameters $N = 25$, $\kappa = 1.5$ and $\delta = 0$

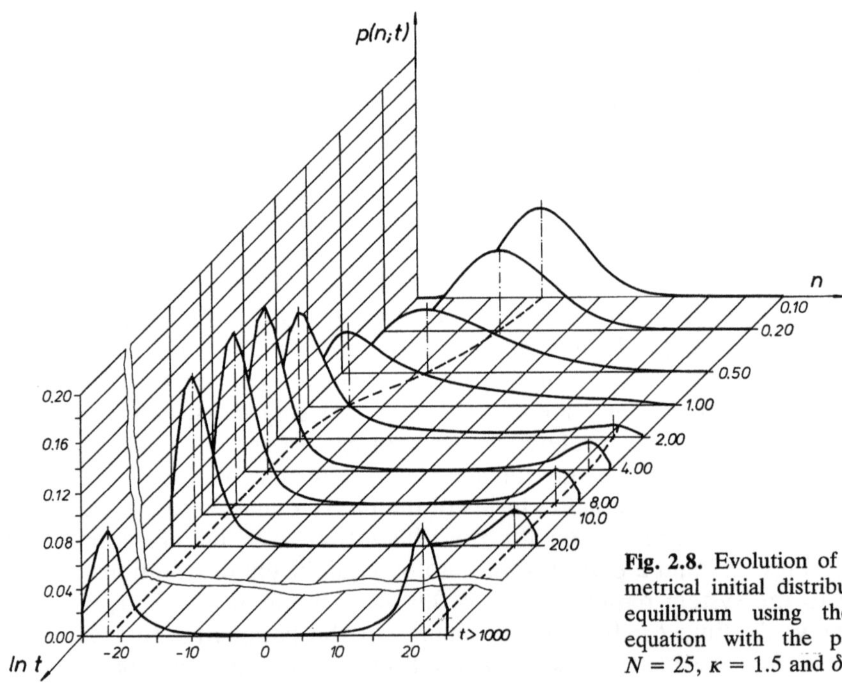

Fig. 2.8. Evolution of an asymmetrical initial distribution into equilibrium using the master equation with the parameters $N = 25$, $\kappa = 1.5$ and $\delta = 0$

the distributions into their new equilibrium shapes as calculated from the master equation.

In Fig. 2.7 the distribution remains symmetrical. The initial fluctuation enhancement and the subsequent drift dominated development of the bimodal distribution into equilibrium can be observed. In Fig. 2.8 the first stages of the development are the same as in Fig. 2.7. In accordance to the asymmetric initial distribution, however, it comes to a bimodal distribution with peaks of differing significance. The final development into the symmetrical stationary distribution is fluctuation dominated and extremely slow. Inserting the explicit equations (2.121, 123, 131) into (2.116) the transition time for collective opinion change is obtained as

$$\mathcal{T} = \frac{\tau}{\sqrt{(\kappa-1)|\kappa(1-x_m^2)-1|}} \exp\left[2N\int_0^{x_m} \frac{\tanh(\kappa x')-x'}{1-x'\tanh(\kappa x')}\,dx'\right] \approx 10^3 \qquad (2.135)$$

in scaled time units for $N = 25$, $\kappa = 1.5$ and $\delta = 0$.

Using this equation Fig. 2.9 shows the strong quasi-exponential dependence of $\mathcal{T}$ (for given N) on the adaptation parameter κ. This result can be regarded as evidence that in totalitarian systems ($\kappa > 1$) the strength of the adaptation trend decisively influences the length of the time period in which a collective change of opinion can be considered as being probable.

In Fig. 2.10 the dependence of the transition time $\mathcal{T}$ on N for given adaptation parameters κ is plotted. This could be relevant when e.g. firms of the same industrial branch are considered as "individuals" whose economic decision behaviour interacts collectively (Chap. 5).

Case b). Secondly an extension of the original model will be discussed. Up to now the trend or control parameters have been treated as constant with time.

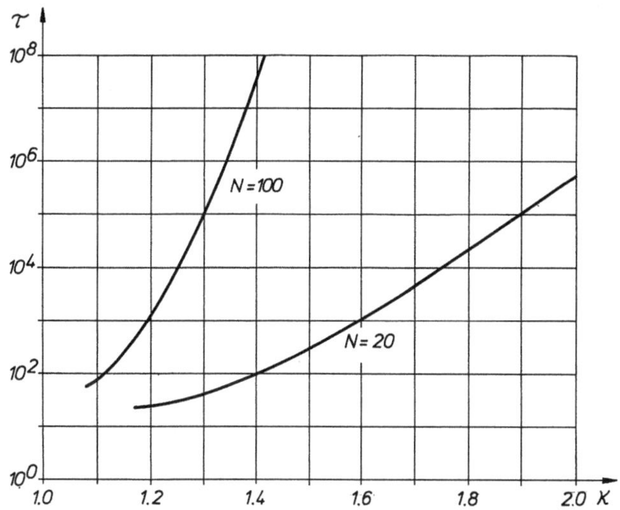

Fig. 2.9. Transition time $\mathcal{T}$ for collective opinion change of a biased opinion distribution into equilibrium wrt the adaptation parameter

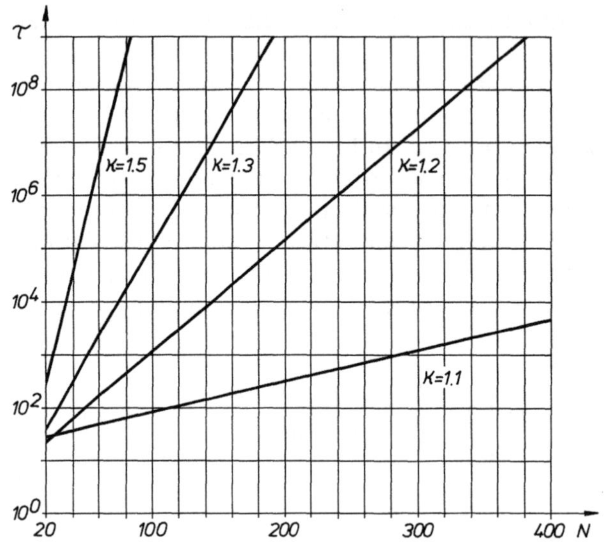

Fig. 2.10. Transition time τ wrt N for given adaptation parameter κ

In reality there may exist a feedback between the socio-configuration and the trend parameters in which case the latter also become *dynamic variables*. This general problem will also be discussed in Chap. 3 and in later models. Here the problem is simplified by assigning a certain *assumed* time dependence

$$\delta(t) = \delta_0 \sin\left(\frac{2\pi}{T}t\right), \tag{2.136}$$

i.e. an alternation with frequency $\omega = 2\pi/T$ and amplitude δ_0 around $\delta = 0$, to the *preference parameter* δ, while keeping κ and ν constant. The time dependence is assigned to δ because this political trend parameter seems to be the one most easily influenced by the given situation and through personal impressions and because periodic changes in preference trends can factually be observed in societies.

The $\delta(t)$ given by (2.136) is inserted into the equations of motion (2.48, 49) for the mean value $\langle x \rangle_{t'}$ and variance $\sigma(t')$. The period T and the path of $\langle x \rangle$ and σ is measured in units of scaled time $t' = \nu t$. Of course, the periodic variation in $\delta(t')$ imposes a variation to $\langle x \rangle_{t'}$ and $\Sigma(t')$ with the same period. However, this is a trivial aspect of the time dependence induced by the assumption (2.136). The dependence of $\langle x \rangle_{t'}$ and $\Sigma(t')$ on T and in particular on κ within the period will now be discussed.

The choice of $\kappa = 0.5$ and $T = 10$ in Fig. 2.11 a corresponds to the situation in a liberal society (small κ) with relatively quick changes in preference trends (small T). The meanvalue $\langle x \rangle_{t'}$ of the opinion configuration follows the alternations of $\delta(t')$ smoothly and closely, though with some delay. The variance and width of the accompanying distribution varies moderately; it is larger when $\langle x \rangle_{t'}$ changes sign and smaller when $|\langle x \rangle_{t'}|$ reaches its maximum.

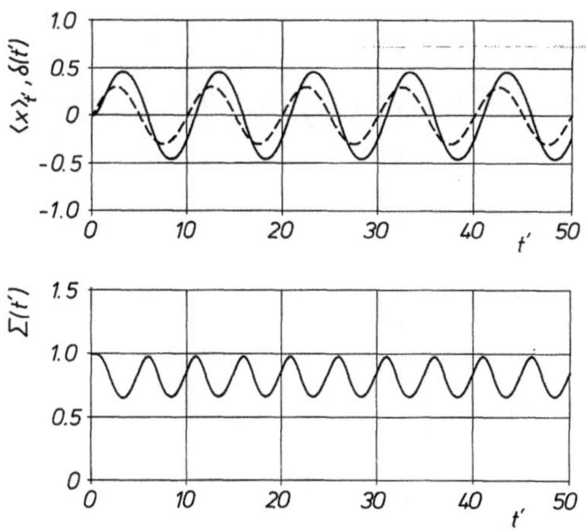

Fig. 2.11 a. Motion of $\langle x \rangle_{t'}$ and $\Sigma(t')$ for $\kappa = 0.5$ and $T = 10$ and for given $\delta(t')$ (---)

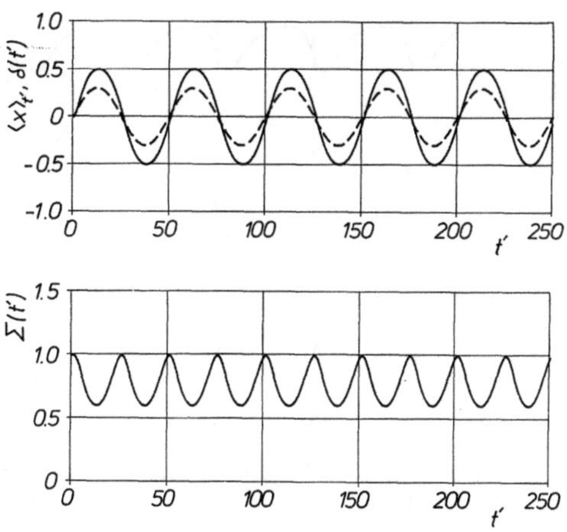

Fig. 2.11 b. Motion of $\langle x \rangle_{t'}$ and $\Sigma(t')$ for $\kappa = 0.5$ and $T = 50$ and for given $\delta(t')$ (---)

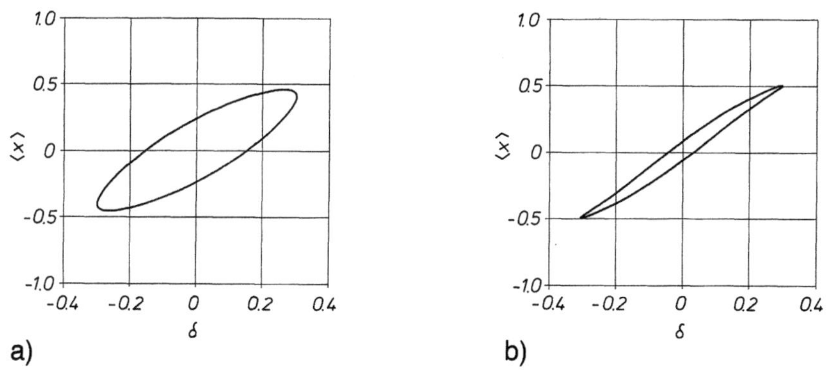

Figs. 2.12 a, b. The mean value $\langle x \rangle$ as function of the preference parameter δ in the cases (**a**) $\kappa = 0.5$, $T = 10$ and (**b**) $\kappa = 0.5$, $T = 50$

The choice of $\kappa = 0.5$ and $T = 50$ in Fig. 2.11 b corresponds to the situation in a liberal society with slow changes in preference trends (large T). The time dependence of all the variables is almost the same as in Fig. 2.11 a but now the opinion configuration $\langle x \rangle_{t'}$ follows the slow variations in $\delta(t')$ almost without delay.

In the Figs. 2.12 a, b time has been eliminated and $\langle x \rangle$ is represented as a function of the preference parameter δ. Illustrated are so-called hysteresis loops for which the Figs. 2.12 a, b correspond to the results of Figs. 2.11 a, b. These figures show that the motion of the socio-configuration (collective opinion configuration) in a liberal society follows closely and in smooth evolutionary way the development of the individual preference trends described by δ. Further, the moderate values and variations in the variance (i.e. the dispersion of the distribution) lead to small fluctuations of sample paths around their mean

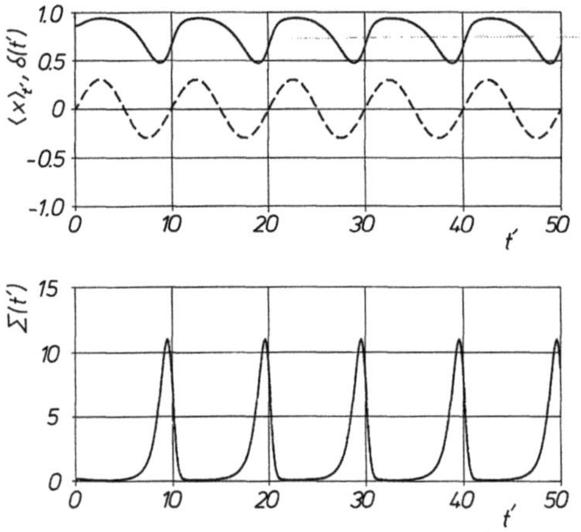

Fig. 2.13 a. Motion of $\langle x \rangle_{t'}$ and $\Sigma(t')$ for $\kappa = 1.5$ and $T = 10$ and for given $\delta(t')$ (---)

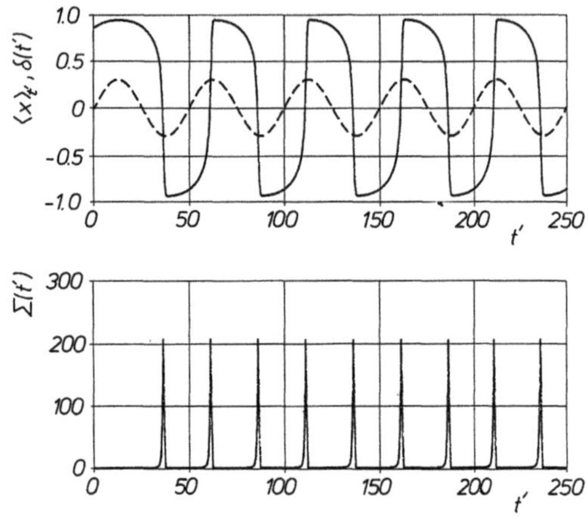

Fig. 2.13 b. Motion of $\langle x \rangle_{t'}$ and $\Sigma(t')$ for $\kappa = 1.5$ and $T = 50$ and for given $\delta(t')$ (---)

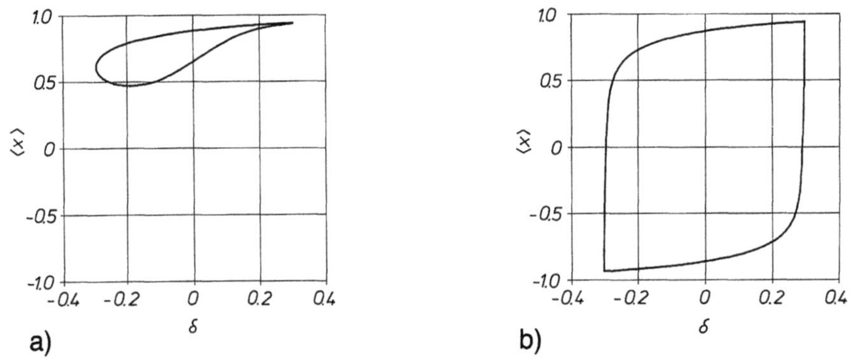

Figs. 2.14 a, b. The mean value $\langle x \rangle$ as function of the preference parameter δ in the cases (**a**) $\kappa = 1.5$, $T = 10$ and (**b**) $\kappa = 1.5$, $T = 50$

values. It can be summarized that liberal societies produce soft, evolutionary and predictable developments in their socio-configuration.

On the other hand, the choice of $\kappa = 1.5$ and $T = 10$ (in Fig. 2.13 a) corresponds to a situation in a totalitarian society (large κ above the critical value) with rapid changes in preference trends $\delta(t')$. Although the individual preference trends develop a large amplitude opposed to the mean value $\langle x \rangle_{t'}$ the quick alternation of $\delta(t')$ prevents the breakdown of the prevalent opinion stabilized by the large adaptation κ. Nevertheless, in periods in which opposite signs for $\delta(t')$ and $\langle x \rangle_{t'}$ occur, a high variance corresponding to large fluctuations develops.

The choice of $\kappa = 1.5$ and $T = 50$ in Fig. 2.13 b corresponds to a situation in a totalitarian society with slow changes in the preference trend $\delta(t')$. The prevalent, adaptation stabilized opinion initially remains stable although the individual preferences have reached a large amplitude of opposite sign. Then a sudden (revolutionary) breakdown of the collective opinion configuration into the opposite direction sets in. This corresponds to a phase transition of the first order as can be seen from the change in the shape of the potential $V(x, \kappa, \delta)$, see (2.122). Because of the large κ the opinion then stabilizes in the opposite direction until the subsequent alternation of the preference trend leads to the next "revolution". While very small variance in the quasi-stationary opinion states occurs, there is a large increase in $\Sigma(t')$ during the revolutionary breakdown. The corresponding high fluctuations cause a loss in predictability for the single sample society during this period.

In Fig. 2.14 $\langle x \rangle$ is again represented as a function of the preference parameter δ. Hysteresis loops of a very different shape are now obtained. The Fig. 2.14a corresponds to Fig. 2.13a and describes a "small revolutionary cycle". In spite of large preference oscillations the adaptation stabilized opinion mean value retains its sign. The Fig. 2.14b corresponds to Fig. 2.13b and describes a "large revolutionary cycle" containing the quasi-stable as well as a sudden breakdown behaviour for $\langle x \rangle$.

Summarizing, the model solutions of Figs. 13 and 14 can be interpreted as suggesting that totalitarian societies produce – even after delusively stable periods – hard, revolutionary and in their details unpredictable transitions in their socio-configuration.

It is left to the reader to apply – and to extend the implications inherent in – the model in making an inital interpretation of many relevant events of recent history, in particular of the smooth evolutionary transitions in liberal societies and the hard revolutionary transitions in totalitarian societies (e.g. the Carter-Reagan change represents a "smooth" transition and the Shah-Ayatollah change a "hard" transition).

3. Fundamental Concepts of Quantitative Sociology

In this chapter the concepts of a quantitative approach to dynamic processes in socio-systems will be introduced in general form. The procedure consists of two steps: First of all appropriate macrovariables of the socio-system have to be found. Secondly, equations of motion for these macrovariables must be established.

3.1 Attitude Space, Socio-Configuration and Situation Space

In the Introduction it has been already explained that it is impossible – due to the lack of detailed information and analysis – to develop a *microscopic* theory of sociological processes. Interest therefore is confined to the observation of the time paths of certain *macrostates* using a phenomenological probabilistic description.

Nevertheless a beginning has to be made by observing the individual members of a society. There exist different and distinguishable *"aspects of life"* in which the individual necessarily has to assume his role. The various aspects of politics, religion, education, consumer behaviour, etc. have already been mentioned. Although such aspects can be effectively delineated, the attitudes of individuals to them can clearly be interrelated (such interrelationships would – once detected and modelled – be reflected in the equations of motion).

The delineation and definition of aspects should be chosen in such a way as to allow for a manifold of different *attitudes* (for instance, opinions, denominations, qualifications, buying habits, etc.) with respect to the aspects mentioned above. From the standpoint of the dynamics of their origin, the cells of the brain, attitudes are already macrovariable states belonging to the individual, i.e. a resultant projection of complex thoughts, emotions, experiences and actions.

In order to avoid problems of topology, in particular the problem of a metrically defined "distance" in attitude space, it is preferred to attribute a *discrete number* of possible attitudes to each aspect.

Let it be assumed that A aspects exist and attribute d_a possible attitudes $i_a = 1, 2, \ldots, d_a$ to each aspect a with $a = 1, 2, \ldots, A$. It is then convenient to introduce the *attitude space* $\mathfrak{A}$ spanned by A dimensions belonging to the A different aspects. The *attitude vector* $\boldsymbol{i} = \{i_1, i_2, \ldots, i_A\}$ of an individual then points to a lattice point in $\mathfrak{A}$ and characterizes the attitude of the individual with respect to all A aspects. By assumption there exist

$$L = \prod_{a=1}^{A} d_a \tag{3.1}$$

such lattice points.

If a society consisting of N members is now considered, it can be subdivided into P subpopulations $\mathcal{P}_\alpha$, $(\alpha = 1, 2, \ldots, P)$ each with N_α members where

$$N = \sum_{\alpha=1}^{P} N_\alpha \tag{3.2}$$

if this is indicated as justifiable and relevant by the distinguishable probabilistic behaviour of members of the different subpopulations with respect to the aspects in $\mathfrak{A}$.

In general birth and death and/or immigration and emigration processes will be included; therefore the N_α need not to be constant with time.

The N_α members of a subpopulation $\mathcal{P}_\alpha$ in general will have different attitudes. Defining $n_{\alpha i}$ to be the number of members of $\mathcal{P}_\alpha$ having the attitude $\boldsymbol{i} = \{i_1, i_2, \ldots, i_A\}$ then:

$$N_i = \sum_{\alpha=1}^{P} n_{\alpha i} \tag{3.3}$$

are all members of the society of attitude $\boldsymbol{i}$ and the following trivial relations

$$N_\alpha = \sum_{i=1}^{L} n_{\alpha i} \quad \text{and} \quad N = \sum_{i=1}^{L}\sum_{\alpha=1}^{P} n_{\alpha i} = \sum_{i=1}^{L} N_i = \sum_{\alpha=1}^{P} N_\alpha \tag{3.4}$$

are easily seen to hold good.

The *macrostate of the society* is now described by the *socio-configuration*

$$\boldsymbol{n} = \{n_{\alpha i}\} \equiv \{n_{11}, \ldots, n_{P1}, \ldots, n_{\alpha i}, \ldots, n_{1L}, \ldots, n_{PL}\} \tag{3.5}$$

consisting of the $C = PL$ integer elements $n_{\alpha i} \geq 0$. The socio-configuration describes the distribution of attitudes among the members of the subpopulations $\mathcal{P}_\alpha$ in the society.

Introducing the C-dimensional *socio-configuration space* $\mathfrak{C}$ the time dependent socio-configuration of a sample society can be represented by a moving point $\boldsymbol{n}(t) = \{n_{\alpha i}(t)\}$ in the space $\mathfrak{C}$. This space $\mathfrak{C}$ plays a similar role to that of the *phase space* Σ *for macrovariables* in statistical physics. Under the assumption $n_{\alpha i} \gg 1$ one can, of course, go over to quasi-continuous variables $x_{\alpha i} = n_{\alpha i}/N_0$ with appropriately chosen $N_0 \gg 1$.

The attitudes and the socio-configuration, which are indispensible for an understanding of a society's internal dynamics, are mainly related to the psychology and activities of the individuals and determine the cultural and political atmosphere of a society. In addition to the *psychological variables*, variables describing the *material situation* of the society have to be taken into account; for example, those variables which are the subject of *theoretical economics*, like the gross national product, production and investment rates, inflation rates, the employment situation, capital cumulation, innovation intensity, etc. could be used to describe the material situation. These variables seem to be relatively easy to quantify because they can be traced back to material values and valuations. But already here there arise difficult problems connected with the appropriate choice of quantifying parameters.

There exist, however, other more complex components of the "material situation" which are related to the dimensions of the aspect space and whose quantification encounters even more difficulties. For example: The degree of efficiency or bureaucracy in an administration; the degree of "capitalistic" or "socialistic" structures in the economy of a society; the degree of liberal or authoritarian influences in the legislation of a society. In such cases the general method of quantification can only (reasonably) consist in the introduction of a weighted average over the possible realizations of an appropriate set of indicators.

After the definition of quantitative measures y_β ($\beta = 1, 2, \ldots, S$) for the state of S aspects of the material situation, an S-dimensional *situation space* $\mathfrak{S}$ can be introduced and the situation can be described formally by the *situation vector*:

$$y = \{y_1, y_2, \ldots, y_s\} \in \mathfrak{S} \tag{3.6}$$

In terms of the concepts of "socio-configuration" and "situation vector" applied here the dynamics of the society to be described is generated as follows:

For a given situation vector $\boldsymbol{y}$ and under the influence of the inherent *trend parameters* $\boldsymbol{\kappa} = \{\kappa_1, \ldots, \kappa_T\}$ of the psychological constitution of the individuals, there develops a certain attitude distribution or socio-configuration $\boldsymbol{n}(t)$. The motion of $\boldsymbol{n}(t)$ is due to transitions between attitudes caused by individual decision processes. On the other hand, a society with a given socio-configuration $\boldsymbol{n}(t)$ influences and drives the situation vector into new states $\boldsymbol{y}(t)$. For instance, people with certain attitudes develop corresponding activities which change the economic and legislative situation. The changed situation $\boldsymbol{y}(t)$ will then in turn react on the socio-configuration $\boldsymbol{n}(t)$. Under simple assumptions the motion may end in a stable stationary state $\{\boldsymbol{n}_{st}, \boldsymbol{y}_{st}\}$. In general, however, the interactions between $\boldsymbol{n}(t)$ and $\boldsymbol{y}(t)$ will lead to a perpetuated highly complex coupled motion of $\boldsymbol{n}(t)$ and $\boldsymbol{y}(t)$. The models in this book give several examples for this case.

The definitions of socio-configuration and situation vector so far completely leave open *which* aspects of individual attitudes and *which* aspects of the material situation have to be taken into account. Of course, it cannot be expected that *all* individual and situational aspects can be treated simultaneously. It is initially a sufficient achievement to be able to describe the *subdynamics* of a sector of a given society represented in a *subspace* of $\mathfrak{A}$ and $\mathfrak{S}$.

The main problem of such an analysis consists in the *selection* of a set of individual and situational aspects whose dynamics can be *separated* and *decoupled* from the rest of the variables over the considered period of time. The quality and validity of a model depends on this possibility of selecting a few variables with self-contained dynamics.

3.2 Equations of Motion for the Socio-Configuration

It has been explained in Sect. 3.1, that the motions of the socio-configuration and the situation vector are coupled. Nevertheless it is reasonable in a first approximation to treat the equations for the socio-configuration $\boldsymbol{n}(t)$ separately. The justification for this is that the motion of $\boldsymbol{n}(t)$ is due to collective psychological processes which in general are much more flexible than the material situation $\boldsymbol{y}(t)$. This implies that $\boldsymbol{n}(t)$ often (but not always) changes much more quickly than $\boldsymbol{y}(t)$. In these cases $\boldsymbol{y}(t)$ can be regarded as quasi-constant and the coupling of its dynamics with $\boldsymbol{n}(t)$ need not be considered. This is the approximation adopted in the models of Chapters 2, 4.

3.2.1 The Master Equation

The most fundamental description of the dynamics of the socio-configuration (3.5) including its fluctuations is given by a master equation for its probability distribution.

First the probability

$$p(\{n_{ai}\};\, t) \equiv p(\boldsymbol{n};\, t) \tag{3.7}$$

is introduced for the socio-configuration to be in state $\{n_{ai}\}$ at time t. The normalization condition is

$$\sum_{\{\boldsymbol{n}\}} p(\boldsymbol{n};\, t) = 1 \tag{3.8}$$

where the sum extends over all possible multiples of non-negative integers $\{n_{ai}\}$, i.e. over $0 \le n_{ai} < \infty$.

Next, the so-called *conditional probability*

$$p_{t+\tau,\,t}(\boldsymbol{n}_0 + \boldsymbol{k},\, \boldsymbol{n}_0) \tag{3.9}$$

is introduced which is the probability of finding a configuration

$\boldsymbol{n}_0 + \boldsymbol{k} \equiv \{n_{011} + k_{11}, \ldots, n_{0ai} + k_{ai}, \ldots, n_{0PL} + k_{PL}\}$

at time $(t + \tau) \geq t$ with $\boldsymbol{k} \equiv \{k_{ai}\}, \, k_{ai} = 0, \pm 1, \pm 2, \pm 3, \ldots$

under the assumption of an initial configuration

$\boldsymbol{n}_0 \equiv \{n_{011}, \ldots, n_{0ia}, \ldots, n_{0PL}\}$ at time t.

In other words, (3.9) is the probability distribution $p_{\boldsymbol{n}_0;\, t}(\boldsymbol{n}; t + \tau)$ with $\boldsymbol{n} = (\boldsymbol{n}_0 + \boldsymbol{k})$ starting from an initial distribution

$p(\boldsymbol{n}; t) = \delta_{\boldsymbol{n}_0 \boldsymbol{n}}$, where $\delta_{\boldsymbol{n}_0 \boldsymbol{n}}$ is the Kronecker symbol.

Obviously this initial distribution is concentrated at the point $\boldsymbol{n} = \boldsymbol{n}_0$. The Markov assumption is now made that the conditional probability does *not depend* on the motion *before* time t which led to the initial configuration $\boldsymbol{n}_0$. The expression (3.9) is zero by definition if $\boldsymbol{n}_0 + \boldsymbol{k}$ or $\boldsymbol{n}_0$ contain negative numbers $n_{0ai} + k_{ai}$ or n_{0ai}, as these are invalid configurations.

Since one of the (valid) configurations $\boldsymbol{n} = \boldsymbol{n}_0 + \boldsymbol{k}$ will *always* be attained at time $t + \tau$ when starting from any arbitrary (valid) configuration $\boldsymbol{n}_0$,

$$\sum_{\{\boldsymbol{k}\}} p_{t+\tau,\, t}(\boldsymbol{n}_0 + \boldsymbol{k}, \boldsymbol{n}_0) \equiv \sum_{\{\boldsymbol{n}\}} p_{\boldsymbol{n}_0;\, t}(\boldsymbol{n}; t + \tau) = 1 \tag{3.10}$$

holds, where the sum over $\{\boldsymbol{k}\}$ extends over all multiples of integers $\{k_{ai}\}$ where $-\infty < k_{ai} < +\infty$.

Further, since any probability $p(\boldsymbol{n}; t + \tau)$ is generated by transitions from all configurations $\boldsymbol{n} + \boldsymbol{k}$ occupied with probabilities $p(\boldsymbol{n} + \boldsymbol{k}; t)$ at time t, it follows by definition of the conditional probability that

$$p(\boldsymbol{n}; t + \tau) = \sum_{\{\boldsymbol{k}\}} p_{t+\tau,\, t}(\boldsymbol{n}, \boldsymbol{n} + \boldsymbol{k})\, p(\boldsymbol{n} + \boldsymbol{k}; t) \,. \tag{3.11}$$

Equation (3.11) determines the evolution with time of the probability distribution $p(\boldsymbol{n}; t)$ because it connects $p(\boldsymbol{n}'; t)$ with $p(\boldsymbol{n}; t + \tau)$.

The decisive assumption is now made that for infinitesimally small τ the conditional probability (3.9) for $\boldsymbol{k} \neq \boldsymbol{0}$ is proportional to τ:

$$\lim_{\tau \to 0} p_{t+\tau,\, t}(\boldsymbol{n}_0 + \boldsymbol{k}, \boldsymbol{n}_0) = \tau\, w_t(\boldsymbol{n}_0 + \boldsymbol{k}, \boldsymbol{n}_0) \equiv \tau\, w_t[\boldsymbol{k}, \boldsymbol{n}_0] \tag{3.12}$$

while for $\boldsymbol{k} = \boldsymbol{0}$ it follows from (3.10, 12) that

$$\lim_{\tau \to 0} p_{t+\tau,\, t}(\boldsymbol{n}_0, \boldsymbol{n}_0) = 1 - \sum_{\{\boldsymbol{k}\} \neq \{\boldsymbol{0}\}} \tau\, w_t(\boldsymbol{n}_0 + \boldsymbol{k}, \boldsymbol{n}_0) \,. \tag{3.13}$$

$w_t(\boldsymbol{n}_0 + \boldsymbol{k}, \boldsymbol{n}_0)$ is denoted as the *transition probability per unit time* for a configuration change from $\boldsymbol{n}_0$ to $\boldsymbol{n}_0 + \boldsymbol{k}$.

Inserting (3.12, 13) into (3.11) and generalizing (2.9) *the fundamental master equation* corresponding to (1.16) is obtained as:

$$\begin{aligned}\frac{dp(\boldsymbol{n};t)}{dt} &= \lim_{\tau\to 0}\frac{p(\boldsymbol{n};t+\tau)-p(\boldsymbol{n};t)}{\tau}\\ &= \sum_{\{\boldsymbol{k}\}}\{w_t(\boldsymbol{n},\boldsymbol{n}+\boldsymbol{k})p(\boldsymbol{n}+\boldsymbol{k};t)-w_t(\boldsymbol{n}+\boldsymbol{k},\boldsymbol{n})p(\boldsymbol{n};t)\}\\ &\equiv \sum_{\{\boldsymbol{k}\}}\{w_t[-\boldsymbol{k},\boldsymbol{n}+\boldsymbol{k}]p(\boldsymbol{n}+\boldsymbol{k};t)-w_t[-\boldsymbol{k},\boldsymbol{n}]p(\boldsymbol{n};t)\}\ .\end{aligned} \tag{3.14}$$

The last line follows by going over to the notation $w_t(\boldsymbol{n}+\boldsymbol{k},\boldsymbol{n}) = w_t[\boldsymbol{k},\boldsymbol{n}]$, see (3.12), and by substituting $\boldsymbol{k} = -\boldsymbol{k}$ in the last term. The sum extends over all $\boldsymbol{k} = \{k_{\alpha i}\}$.

It must now be considered how the transition probability between socio-configurations introduced in (3.12) is built up of transition probabilities on the level of individuals. The most important transitions are those when one member of the population $\mathcal{P}_\alpha$ or simultaneously g members of a group (for instance, a family) of $\mathcal{P}_\alpha$ change from attitude $\boldsymbol{i}$ to attitude $\boldsymbol{j}$. These individual transition probabilities per time unit are denoted by

$$p_{ji}^{\alpha g}(\boldsymbol{n},\boldsymbol{\kappa})\ . \tag{3.15}$$

In general they depend on the socio-configuration $\boldsymbol{n}$ and on trend parameters $\boldsymbol{\kappa}$ (e.g. the adaptation and preference parameters of Chap. 2). The latter are considered as constants here, but may also become dynamic variables in an extended treatment (see, for instance, Chap. 5).

As there exist $n_{\alpha i}/g$ groups (families) of g members with attitude $\boldsymbol{i}$ in $\mathcal{P}_\alpha$, the contribution of (3.15) to $w_t[\boldsymbol{k},\boldsymbol{n}]$ is

$$\left.\begin{aligned} w_{ji}^{\alpha g}[\boldsymbol{k},\boldsymbol{n}] &= \frac{n_{\alpha i}}{g}\,p_{ji}^{\alpha g}(\boldsymbol{n},\boldsymbol{\kappa}) \quad \text{for}\quad \boldsymbol{k} = \{0\dots g_{\alpha j}\dots 0\dots -g_{\alpha i}\dots 0\dots\}\\ w_{ji}^{\alpha g}[\boldsymbol{k},\boldsymbol{n}] &= 0 \qquad \text{for all other}\quad \boldsymbol{k}\ .\end{aligned}\right\} \tag{3.16}$$

The vector $\boldsymbol{k}$ for non-vanishing transition probabilities $w_{ji}^{\alpha g}$ contains zeros except for the integers g and $(-g)$ in the positions αj *and* αi. *The transitions (3.16) conserve the numbers* N_α.

Another kind of transition occurs if a group of g members of one subpopulation $\mathcal{P}_\alpha$ (for instance of a given educational or economic level) changes per definitionem into another population $\mathcal{P}_\beta$ but keeps the previous attitude $\boldsymbol{i}$. The corresponding individual transition probability

$$p_{ig}^{\beta\alpha}(\boldsymbol{n},\boldsymbol{\kappa}) \tag{3.17}$$

then contributes to $w_t[\boldsymbol{k},\boldsymbol{n}]$ as follows:

$$\left.\begin{aligned} w_{ig}^{\beta\alpha}[\boldsymbol{k}, \boldsymbol{n}] &= \frac{n_{\alpha i}}{g} p_{ig}^{\beta\alpha}(\boldsymbol{n}, \boldsymbol{\kappa}) \qquad \text{for} \quad \boldsymbol{k} = \{\ldots 0 \ldots g_{\beta i} \ldots 0 \ldots -g_{\alpha i} \ldots 0 \ldots\} \\ w_{ig}^{\beta\alpha}[\boldsymbol{k}, \boldsymbol{n}] &= 0 \qquad \text{for all other} \quad \boldsymbol{k}\,. \end{aligned}\right\} \tag{3.18}$$

The vector $\boldsymbol{k}$ for non-vanishing transition probabilities $w_{ig}^{\beta\alpha}$ contains zeros except for the integers g and $(-g)$ in the positions βi and αi. The transitions (3.18) conserve the numbers N_i.

Finally birth and death processes and/or immigration and emigration processes are considered. The individual probabilities per time unit for birth (immigration) and death (emigration) of g members of $\mathcal{P}_\alpha$ with attitude $\boldsymbol{i}$ are denoted by

$$p_{i+}^{\alpha g}(\boldsymbol{n}, \boldsymbol{\kappa}) \qquad \text{and} \quad p_{i-}^{\alpha g}(\boldsymbol{n}, \boldsymbol{\kappa}) \tag{3.19}$$

and contribute to $w_t[\boldsymbol{k}, \boldsymbol{n}]$ as follows:

$$\left.\begin{aligned} w_{i+}^{\alpha g}[\boldsymbol{k}, \boldsymbol{n}] &= m_{\alpha i} p_{i+}^{\alpha g}(\boldsymbol{n}, \boldsymbol{\kappa}) \qquad \text{for} \quad \boldsymbol{k} = \{\ldots 0 \ldots g_{\alpha i} \ldots 0 \ldots\} \\ w_{i+}^{\alpha g}[\boldsymbol{k}, \boldsymbol{n}] &= 0 \qquad \text{for all other} \quad \boldsymbol{k} \end{aligned}\right\} \tag{3.20 a}$$

and

$$\left.\begin{aligned} w_{i-}^{\alpha g}[\boldsymbol{k}, \boldsymbol{n}] &= n_{\alpha i} p_{i-}^{\alpha g}(\boldsymbol{n}, \boldsymbol{\kappa}) \qquad \text{for} \quad \boldsymbol{k} = \{\ldots 0 \ldots -g_{\alpha i} \ldots 0 \ldots\} \\ w_{i-}^{\alpha g}[\boldsymbol{k}, \boldsymbol{n}] &= 0 \qquad \text{for all other} \quad \boldsymbol{k}\,. \end{aligned}\right\} \tag{3.20 b}$$

For birth processes $m_{\alpha i}$ will be proportional to the number of parents ($\sim n_{\alpha i}$); in the case of immigration $m_{\alpha i}$ is the number of potential immigrants of the kind αi. The vectors $\boldsymbol{k}$ for non-vanishing transition probabilities $w_{i+}^{\alpha g}$ and $w_{i-}^{\alpha g}$ contain zeros except for g and $(-g)$ in the respective positions αi. Neither N_i nor N_α are conserved for birth and death processes, instead the society and its subpopulations can grow or shrink. It depends on the model being considered whether $w_t[\boldsymbol{k}, \boldsymbol{n}]$ is a sum of *all* the contributions (3.16–20) or whether some of them can be neglected.

In the general case the composition formula for the transition probability between socio-configurations is:

$$\begin{aligned} w_t[\boldsymbol{k}, \boldsymbol{n}] = &\sum_{j,i,g} \sum_{\alpha} w_{ji}^{\alpha g}[\boldsymbol{k}, \boldsymbol{n}] \\ &+ \sum_{\beta,\alpha,g} \sum_{i} w_{ig}^{\beta\alpha}[\boldsymbol{k}, \boldsymbol{n}] \\ &+ \sum_{i,\alpha,g} \{w_{i+}^{\alpha g}[\boldsymbol{k}, \boldsymbol{n}] + w_{i-}^{\alpha g}[\boldsymbol{k}, \boldsymbol{n}]\} \end{aligned} \tag{3.21}$$

where the index t indicates of the possibility that the transition probabilities themselves may still be time dependent, for instance when the trend parameters are dynamic variables (see Sect. 3.3).

3.2.2 The Stochastic and the Fokker-Planck Equation

Starting from the master equation (3.14) the *stochastic equation* and subsequently the *Fokker-Planck equation* – using the *Kramers Moyal expansion* – can be derived in a general and straight forward manner.

From now $p(\boldsymbol{n}; t)$ and $w_t[\boldsymbol{k}, \boldsymbol{n}]$ are considered as *differentiable functions* of the C variables $\boldsymbol{n} = \{n_{ai}\}$. Expanding the first term on the rhs of (3.14) in a Taylor series with respect to $\boldsymbol{n} + \boldsymbol{k}$ around $\boldsymbol{n}$:

$$\{w_t[-\boldsymbol{k}, \boldsymbol{n} + \boldsymbol{k}]p(\boldsymbol{n} + \boldsymbol{k}; t) - w_t[-\boldsymbol{k}, \boldsymbol{n}]p(\boldsymbol{n}; t)\}$$

$$= \sum_{s_1=1}^{\infty} \dots \sum_{s_c=1}^{\infty} \frac{k_1^{s_1} \dots k_c^{s_c}}{s_1! \dots s_c!} \cdot \frac{\partial^{s_1}}{\partial n_1^{s_1}} \dots \frac{\partial^{s_c}}{\partial n_c^{s_c}} (w_t[-\boldsymbol{k}, \boldsymbol{n}]p(\boldsymbol{n}; t)) \tag{3.22}$$

is obtained. Finally, by introducing the *moments* of order $s = s_1 + s_2 + \dots + s_c$:

$$\begin{aligned} m_{s_1 \dots s_c}(\boldsymbol{n}_0; t) &= \sum_{\{\boldsymbol{k}\}} k_1^{s_1} \dots k_c^{s_c} w_t[\boldsymbol{k}, \boldsymbol{n}_0] \\ &= \lim_{\tau \to 0} \frac{1}{\tau} \sum_{\{\boldsymbol{k}\}} k_1^{s_1} \dots k_c^{s_c} p_{t+\tau, t}(\boldsymbol{n}_0 + \boldsymbol{k}, \boldsymbol{n}_0) \\ &= \lim_{\tau \to 0} \frac{1}{\tau} \langle [n_1(t+\tau) - n_{01}(t)]^{s_1} \dots [n_c(t+\tau) - n_{0c}(t)]^{s_c} \rangle \end{aligned} \tag{3.23}$$

and inserting (3.22) with (3.23) into (3.14), the stochastic equation

$$\frac{\partial p(\boldsymbol{n}; t)}{\partial t} = \sum_{s=1}^{\infty} \frac{(-1)^s}{s!} \sum_{s_1, s_2, \dots, s_c}^{s_1 + \dots s_c = s} \frac{s!}{s_1! \dots s_c!} \frac{\partial^{s_1}}{\partial n_1^{s_1}} \dots \frac{\partial^{s_c}}{\partial n_c^{s_c}} (m_{s_1 \dots s_c}(\boldsymbol{n}; t) p(\boldsymbol{n}; t)) \tag{3.24}$$

is derived. It is convenient to convert to the scaled variables:

$$x_{ai} = \frac{n_{ai}}{N_0} \quad \text{and} \quad \Delta x_{ai} = \frac{1}{N_0} \equiv \varepsilon \tag{3.25}$$

where N_0 may be chosen as the total number of members of the stationary (or quasi-stationary) society.

Simultaneously, the probability distribution over the C quasi-continuous variables $\boldsymbol{x} = \{x_{ai}\}$

$$P(\boldsymbol{x}; t) = N_0^C p(\boldsymbol{n}; t) \tag{3.26}$$

is introduced and normalized to

$$\sum_{\{n\}} p(\boldsymbol{n}; t) = \sum_{\{x\}} P(\boldsymbol{x}; t)\, \Delta x_1 \ldots \Delta x_c \approx \int_0^{\infty} P(\boldsymbol{x}; t)\, d^c \boldsymbol{x} = 1\,. \tag{3.27}$$

Further, because $w_t[\boldsymbol{k}, \boldsymbol{n}]$ is essentially composed of terms proportional to $n_{ai} = N_0 x_{ai}$, a conversion:

$$m_{s_1 \ldots s_c}(\boldsymbol{n}; t) = N_0 M_{s_1 \ldots s_c}(\boldsymbol{x}; t) \tag{3.28}$$

is indicated. Then the stochastic equation reads

$$\frac{\partial P(\boldsymbol{x}; t)}{\partial t} = \sum_{s=1}^{\infty} \frac{(-1)^s \varepsilon^{s-1}}{s!} \sum_{s_1, \ldots, s_c}^{s_1 + \ldots + s_c = s} \frac{s!}{s_1! \ldots s_c!} \frac{\partial^{s_1}}{\partial x_1^{s_1}} \cdots \frac{\partial^{s_c}}{\partial x_c^{s_c}} [M_{s_1 \ldots s_c}(\boldsymbol{x}; t)\, P(\boldsymbol{x}; t)]\,. \tag{3.29}$$

The *Fokker-Planck equation* then follows from the stochastic equation (3.29) as an *approximate equation* by neglecting all terms with $s \geq 3$, which is justified for $N_0 \gg 1$ or $\varepsilon \ll 1$. It may also happen that the moments (3.28) vanish *exactly* for $s \geq 3$. In this case the Fokker-Planck equation is an *exact equation* (Sect. 3.2.3).

For convenience from now on only *one index i* running from 1 to C instead of ai will be written for the components of the socio-configuration. Further, the following notation for moments of first and second order is introduced:

$$\begin{aligned} K_i(\boldsymbol{x}_0; t) &= \varepsilon \sum_{\{k\}} k_i w_t[\boldsymbol{k}, \boldsymbol{n}_0] \\ &= \lim_{\tau \to 0} \frac{1}{\tau} \langle [x_i(t+\tau) - x_{0i}(t)] \rangle \end{aligned} \tag{3.30}$$

$$\begin{aligned} Q_{ij}(\boldsymbol{x}_0; t) &= \varepsilon \sum_{\{k\}} k_i k_j w_t[\boldsymbol{k}, \boldsymbol{n}_0] \\ &= \lim_{\tau \to 0} \frac{1}{\tau} N_0 \langle [x_i(t+\tau) - x_{0i}(t)][x_j(t+\tau) - x_{0j}(t)] \rangle \end{aligned} \tag{3.31}$$

and ∂_i can be written for $\partial/\partial x_i$. Then *the general Fokker-Planck equation for the socio-configuration* is obtained from (3.29) as

$$\begin{aligned} \frac{\partial P(\boldsymbol{x}; t)}{\partial t} &= -\sum_{i=1}^{C} \partial_i [K_i(\boldsymbol{x}; t)\, P(\boldsymbol{x}; t)] + \frac{\varepsilon}{2} \sum_{i,j=1}^{C} \partial_i \partial_j [Q_{ij}(\boldsymbol{x}; t)\, P(\boldsymbol{x}; t)] \\ &= -\sum_{i=1}^{C} \partial_i I_i(\boldsymbol{x}; t) \end{aligned} \tag{3.32}$$

where

$$I_i(\boldsymbol{x};t) = K_i(\boldsymbol{x};t)\,P(\boldsymbol{x};t) - \frac{\varepsilon}{2}\sum_{j=1}^{c}\partial_j[Q_{ij}(\boldsymbol{x};t)\,P(\boldsymbol{x};t)] \tag{3.33}$$

are the components of the probability current. The coefficients $K_i(\boldsymbol{x};t)$ and $Q_{ij}(\boldsymbol{x};t)$ are, in general, time dependent. Methods for the solution of Fokker-Planck equations with time dependent coefficients have already been developed [3.1].

If the trend parameters $\boldsymbol{\kappa}$ are constants with time, the *drift coefficients* $K_i(\boldsymbol{x})$ and the *fluctuation* coefficients $Q_{ij}(\boldsymbol{x})$ become independent of time. Fokker-Planck equations of the form (3.32) with time independent coefficients are treated in textbooks on statistical physics [1.51–53].

There exists another often useful formulation of the Fokker-Planck equation when it is written in the following form

$$\begin{aligned}\frac{\partial P(\boldsymbol{x};t)}{\partial t} &= \sum_{i,j=1}^{c}\partial_i\,\varepsilon L_{ij}(\boldsymbol{x})\left[\partial_j P(\boldsymbol{x};t) - \frac{P(\boldsymbol{x};t)}{W(\boldsymbol{x})}\,\partial_j W(\boldsymbol{x})\right]\\ &= -\sum_{i=1}^{c}\partial_i J_i(\boldsymbol{x};t)\end{aligned} \tag{3.34}$$

with the current

$$J_i(\boldsymbol{x};t) = \sum_{j=1}^{c}\varepsilon L_{ij}(\boldsymbol{x})\left[\frac{P(\boldsymbol{x};t)}{W(\boldsymbol{x})}\,\partial_j W(\boldsymbol{x}) - \partial_j P(\boldsymbol{x};t)\right] \tag{3.35}$$

where $W(\boldsymbol{x})$ is a positive definite function and $L_{ij}(\boldsymbol{x})$ are so called *transport coefficients*. In the sociological systems discussed here the variables keep their sign under time reversal with the consequence that the transport coefficients form a positive definite symmetrical matrix [3.2]

$$L_{ij}(\boldsymbol{x}) = L_{ji}(\boldsymbol{x})\,. \tag{3.36}$$

Putting the probability distribution $P(\boldsymbol{x};t)$ equal to $W(\boldsymbol{x})$ the rhs of (3.34, 35) vanish. This implies

$$\left.\begin{array}{lll} J_i(\boldsymbol{x};t) \Rightarrow 0 & \text{and} & \dfrac{\partial P(\boldsymbol{x};t)}{\partial t} \Rightarrow 0 \\ \text{for} & & P(\boldsymbol{x};t) \Rightarrow W(\boldsymbol{x}) \end{array}\right\} \tag{3.37}$$

and so the function $W(\boldsymbol{x})$ turns out to be the *stationary solution* of a Fokker-Planck equation of the form (3.34). It is convenient to write $W(\boldsymbol{x})$ as

$$W(\boldsymbol{x}) = \exp[\varepsilon^{-1}\Psi(\boldsymbol{x})]\,. \tag{3.38}$$

The *relation* between the *two versions* (3.32) and (3.34) of the Fokker-Planck equation becomes obvious if the coefficients of $P(\boldsymbol{x}; t)$, $\partial_i P(\boldsymbol{x}; t)$ and $\partial_i \partial_j P(\boldsymbol{x}; t)$ on the rhs of both equations are compared. Equations (3.32, 34) are equivalent if

$$Q_{ij}(\boldsymbol{x}) = Q_{ji}(\boldsymbol{x}) = 2\, L_{ij}(\boldsymbol{x}) \tag{3.39}$$

and

$$K_j(\boldsymbol{x}) = \frac{1}{W(\boldsymbol{x})} \sum_{i=1}^{c} \partial_i (\varepsilon\, L_{ji}(\boldsymbol{x})\, W(\boldsymbol{x})) \tag{3.40}$$

holds. In this case the probability currents also agree:

$$I_i(\boldsymbol{x}; t) = J_i(\boldsymbol{x}; t)\ . \tag{3.41}$$

In other words, the Fokker-Planck equation (3.34) and the current (3.35) adopt the standard forms (3.32, 33) with fluctuation coefficients $Q_{ij}(\boldsymbol{x})$ and drift coefficients $K_i(\boldsymbol{x})$ given by (3.39, 40). Drift- and fluctuation coefficients of the form (3.39, 40), however, do not represent the most general case since they satisfy the so called *condition of detailed balance* [3.2–4]. This condition follows by inserting (3.38, 39) in (3.40):

$$K_j(\boldsymbol{x}) = \frac{\varepsilon}{2} \sum_{i=1}^{c} \partial_i Q_{ji}(\boldsymbol{x}) + \frac{1}{2} \sum_{i=1}^{c} Q_{ji}(\boldsymbol{x})\, \partial_i \Psi(\boldsymbol{x}) \tag{3.42}$$

and solving for $\partial_k \Psi(\boldsymbol{x})$ under the assumption that the matrix $Q_{ji}(\boldsymbol{x})$ can be inverted.

Using

$$\sum_{j=1}^{c} Q_{kj}^{(-1)}(\boldsymbol{x})\, Q_{ji}(\boldsymbol{x}) = \delta_{ki} \tag{3.43}$$

the following is obtained from (3.42):

$$\partial_k \Psi(\boldsymbol{x}) = \sum_{j=1}^{c} Q_{kj}^{(-1)}(\boldsymbol{x}) \left[2\, K_j(\boldsymbol{x}) - \varepsilon \sum_{i=1}^{c} \partial_i Q_{ji}(\boldsymbol{x}) \right] . \tag{3.44}$$

Finally, the application of the relation

$$\partial_l \partial_k \Psi(\boldsymbol{x}) = \partial_k \partial_l \Psi(\boldsymbol{x}) \tag{3.45}$$

to (3.44) yields the condition of detailed balance which is automatically fulfilled by coefficients $K_j(\boldsymbol{x})$ and $Q_{ij}(\boldsymbol{x})$ of the forms (3.39, 40)

$$\partial_l \sum_{j=1}^{C} Q_{kj}^{(-1)}(\boldsymbol{x}) \left[2K_j(\boldsymbol{x}) - \varepsilon \sum_{i=1}^{C} \partial_i Q_{ji}(\boldsymbol{x}) \right]$$

$$= \partial_k \sum_{j=1}^{C} Q_{lj}^{(-1)}(\boldsymbol{x}) \left[2K_j(\boldsymbol{x}) - \varepsilon \sum_{i=1}^{C} \partial_i Q_{ji}(\boldsymbol{x}) \right] . \tag{3.46}$$

Vice versa, the standard Fokker-Planck equation (3.32) can be transformed into the form (3.34) if and only if the coefficients $K_i(\boldsymbol{x})$, $Q_{ij}(\boldsymbol{x})$ satisfy the condition of detailed balance (3.46), which is equivalent to the integrability condition for $\Psi(\boldsymbol{x})$ in (3.44). If (3.46) holds, $\Psi(\boldsymbol{x})$ and $W(\boldsymbol{x})$ can be found by integrating (3.44) and, going through (3.36) to (3.43), the Fokker-Planck equation (3.32) can be cast in the form (3.34). On the other hand, it must be said that (3.32) is more general than (3.34) since the drift- and fluctuation coefficients $K_i(\boldsymbol{x})$ and $Q_{ij}(\boldsymbol{x})$ *may*, but need not satisfy the condition of detailed balance.

In sociological models as well as those in physics and chemistry there exist relevant systems described by Fokker-Planck equations for which detailed balance is *not* fulfilled (Chap. 4).

The characteristic difference between systems *with* and *without* detailed balance is the following: In those *with detailed balance* each component of the stationary probability current vanishes separately, see (3.37, 41)

$$I_i^{\mathrm{st}}(\boldsymbol{x}) = 0, \qquad \text{for} \quad i = 1, 2, \ldots, C$$

In systems *without detailed balance* at least some of the components of the stationary current $I_i^{\mathrm{st}}(\boldsymbol{x})$ must be non-vanishing, i.e.

$$I_i^{\mathrm{st}}(\boldsymbol{x}) \neq 0 \qquad \text{for some} \quad i = 1, 2, \ldots, C . \tag{3.47}$$

Of course, the condition

$$\sum_{i=1}^{C} \partial_i I_i^{\mathrm{st}}(\boldsymbol{x}) = 0 \tag{3.48}$$

for the stationary current, which follows from the stationary version of the Fokker-Planck equation (3.32), has to be fulfilled in *all* cases.

3.2.3 *The Langevin Equations and the Fokker-Planck Equation

It has been shown that the key quantities of the phenomenological approach to quantitative sociology in this book are the individual transition probabilities as introduced by (3.15, 17, 19) on which the transition probabilities (3.16, 18, 20)

---

* This section can be omitted on a first reading.

for the socio-configuration are based. It was therefore a natural mathematical procedure to start in Sect. 3.2.1, 2 with equations of motion for the probability distribution (master equation, stochastic equation and Fokker-Planck equation) as these are most directly linked with the transition probabilities. In Chap. 1, on the other hand, the path $\boldsymbol{x}(t)$ of the socio-configuration of a single sample society was considered which is most naturally described by a *Langevin equation* of the type (1.1) with fluctuating forces obeying (1.4–8).

The *relation* between Langevin equations and the Fokker-Planck equation will now be discussed. As in Chap. 1 the starting point are the *Langevin equations of the variables* $\{x_i(t)\}$ *for the socio-configuration* (and eventually including $\{y_i(t)\}$ for the material situation) of the general form:

$$\frac{dx_i(t)}{dt} = k_i(\boldsymbol{x}) + \sum_{j=1}^{c} g_{ij}(\boldsymbol{x})\,\xi_j(t), \qquad \text{where} \quad i = 1, 2, \ldots, C \tag{3.49}$$

with fluctuating forces $\xi_j(t)$ due to micro-processes which are unknown except for some general assumptions about their properties. In order to characterize these properties the r-fold *joint probabilities* of the fluctuating forces can be introduced for an ensemble of systems described by (3.49) as

$$p_r(\xi_1, \ldots, \xi_r; t_1, \ldots, t_r)\, d\tau(\xi_1) \ldots d\tau(\xi_r) \tag{3.50}$$

for finding the fluctuating force $\boldsymbol{\xi} = (\xi_1 \ldots \xi_c)$ at times t_α $(\alpha = 1, 2, \ldots, r)$ in the intervals $d\tau_\alpha(\xi_\alpha)$ $(\alpha = 1, 2, \ldots, r)$. Clearly, the mean value of a function $f(\xi_1, \ldots, \xi_r)$ taken about $\xi_1, \ldots, \xi_r$ at times $t_1, \ldots, t_r$ is given by

$$\begin{aligned} &\langle f(\xi_1(t_1), \ldots, \xi_r(t_r)\rangle \\ &\quad = \int f(\xi_1, \ldots, \xi_r)\, p_r(\xi_1, \ldots, \xi_r; t_1, \ldots, t_r)\, d\tau_1(\xi_1) \ldots d\tau_r(\xi_r)\,. \end{aligned} \tag{3.51}$$

In particular the *multi-time moments* or *correlation functions* of sth order of the fluctuating forces are defined as

$$\begin{aligned} &\boldsymbol{\mu}_s(t_1, \ldots, t_s) \\ &\quad \equiv \{\mu_s(i_1, \ldots, i_s; t_1 \ldots t_s)\} = \{\langle \xi_{i_1}(t_1) \ldots \xi_{i_s}(t_s)\rangle\} \end{aligned} \tag{3.52}$$

and the *characteristic functions* as:

$$\left.\begin{aligned} &\Theta_r(\boldsymbol{u}_1, \ldots, \boldsymbol{u}_r; t_1, \ldots, t_r) = \langle \exp\{\mathrm{i}[\boldsymbol{u}_1 \boldsymbol{\xi}(t_1) + \ldots + \boldsymbol{u}_r \boldsymbol{\xi}(t_r)]\}\rangle \\ &\text{with} \\ &\boldsymbol{u}_\alpha \boldsymbol{\xi}(t_\alpha) \equiv \sum_{j=1}^{c} u_{\alpha j}\, \xi_j(t_\alpha)\,. \end{aligned}\right\} \tag{3.53}$$

The latter functions may be expanded in two different ways; either as the Taylor series:

$$\left.\begin{aligned}&\Theta_r(u_1, \ldots, u_r; t_1, \ldots, t_r)\\&\quad = 1 + \sum_{s=1}^{\infty} \frac{\mathrm{i}^s}{s!} \sum_{\alpha, \ldots, \omega = 1}^{r} \boldsymbol{\mu}_s(t_\alpha, \ldots, t_\omega)\, \boldsymbol{u}_\alpha \ldots \boldsymbol{u}_\omega\\&\text{with}\\&\boldsymbol{\mu}_s(t_\alpha, \ldots, t_\omega)\, \boldsymbol{u}_\alpha \ldots \boldsymbol{u}_\omega \equiv \sum_{i_1 \ldots i_s}^{C} \mu(i_1, \ldots, i_s; t_\alpha, \ldots, t_\omega)\, u_{\alpha i_1} \ldots u_{\omega i_s}\end{aligned}\right\} \tag{3.54}$$

or in the form

$$\left.\begin{aligned}&\Theta_r(u_1, \ldots, u_r; t_1, \ldots, t_r)\\&\quad = \exp\left\{\sum_{s=1}^{\infty} \frac{\mathrm{i}^s}{s!} \sum_{\alpha, \ldots, \omega = 1}^{r} \boldsymbol{\kappa}_s(t_\alpha, \ldots, t_\omega)\, \boldsymbol{u}_\alpha \ldots \boldsymbol{u}_\omega\right\}\\&\text{with}\\&\boldsymbol{\kappa}_s(t_\alpha \ldots t_\omega)\, \boldsymbol{u}_\alpha \ldots \boldsymbol{u}_\omega \equiv \sum_{i_1 \ldots i_s}^{C} \kappa_s(i_1, \ldots, i_s; t_\alpha \ldots t_\omega)\, \boldsymbol{u}_{\alpha i_1} \ldots \boldsymbol{u}_{\omega i_s}\end{aligned}\right\} \tag{3.55}$$

The expansion (3.55) simultaneously defines the *cumulants* of sth order

$$\boldsymbol{\kappa}_s(t_1, \ldots, t_s) = \{\kappa_s(i_1, \ldots, i_s; t_1, \ldots, t_s)\}\ . \tag{3.56}$$

Comparing (3.54) and (3.55) it can easily be checked that the moments (3.52) of sth order can be expressed by the cumulants of first to sth order and vice versa. In particular, the first and second-order moments and cumulants are related as follows:

$$\kappa_1(i; t) = \mu_1(i; t) \equiv \langle \xi_i(t) \rangle \tag{3.57}$$

$$\begin{aligned}\kappa_2(i_1, i_2; t_1, t_2) &= \mu_2(i_1, i_2; t_1, t_2) - \mu_1(i_1; t_1)\mu_1(i_2; t_2)\\&\equiv \langle [\xi_{i_1}(t_1) - \langle \xi_{i_1}(t_1) \rangle][\xi_{i_2}(t_2) - \langle \xi_{i_2}(t_2) \rangle] \rangle\ .\end{aligned} \tag{3.58}$$

Fluctuating forces $\xi_j(t)$ are now assumed with the property that all cumulants vanish except for $\kappa_2(t_1, t_2)$:

$$\begin{aligned}&\kappa_1(i; t) \equiv \langle \xi_i(t) \rangle = 0\\&\kappa_s(i_1, \ldots, i_s; t_1, \ldots, t_s) = 0 \qquad \text{for} \quad s = 3, 4, \ldots, \infty\end{aligned} \tag{3.59}$$

The characteristic functions (3.55) then take the form of Gauss functions:

$$\Theta_r(\boldsymbol{u}_1, \ldots, \boldsymbol{u}_r; t_1, \ldots, t_r) = \exp\left[-\frac{1}{2} \sum_{\alpha,\beta=1}^{r} \boldsymbol{\kappa}_2(t_\alpha, t_\beta)\, \boldsymbol{u}_\alpha \boldsymbol{u}_\beta\right] . \tag{3.60}$$

Equations (3.51, 53) show that $p_r(\boldsymbol{\xi}_1, \ldots, \boldsymbol{\xi}_r; t_1, \ldots, t_r)$ and $\Theta_r(\boldsymbol{u}_1, \ldots, \boldsymbol{u}_r; t_1, \ldots, t_r)$ are Fourier transforms of each other. Since Fourier transforms of Gauss functions are also Gauss functions, multi-variate Gaussian joint probabilities $p_r(\boldsymbol{\xi}_1, \ldots, \boldsymbol{\xi}_r; t_1, \ldots, t_r)$ are then obtained. Fluctuating forces $\xi_j(t)$ fulfilling (3.59) are therefore denoted as a *"Gauss process"* or as Gaussian fluctuating forces.

If, in addition, the second-order cumulant has the form

$$\kappa_2(i, j; t_1, t_2) = \delta_{ij}\delta(t_1 - t_2) \tag{3.61}$$

the $\xi_j(t)$ are denoted as *δ-correlated Gaussian Fluctuating (or random) forces*. Equation (3.61) implies that there are no correlations at all between random forces at *different* times $t_1 \neq t_2$ and also no correlations between the different components $\xi_i(t)$, $\xi_j(t)$ of a random force. In many cases the assumptions (3.59, 61) can be substantiated at least approximately by consideration of the microscopic dynamics producing the fluctuations of $\xi_j(t)$.

The relation between the Langevin equations (3.49) [assuming δ-correlated gaussian random forces obeying (3.59, 61)] and the stochastic equation (3.29) has now to be established. The constitutive quantities of the latter equation are the moments, see (3.23, 28),

$$\begin{aligned} \varepsilon^{s-1} M_{s_1 \ldots s_c}(\boldsymbol{x}_0; t) &= \varepsilon^s m_{s_1 \ldots s_c}(\boldsymbol{n}_0; t) \\ &= \lim_{\tau \to 0} \frac{1}{\tau} \langle \Delta_1^{s_1}(\tau) \ldots \Delta_c^{s_c}(\tau) \rangle \quad \text{with} \quad s = s_1 + \ldots + s_c \end{aligned} \tag{3.62}$$

where the abbreviation

$$\Delta_i(\tau) = [x_i(t + \tau) - x_{0i}(t)] \tag{3.63}$$

has been introduced. These moments will now be calculated, not starting from transition probabilities as in (3.23), but from the Langevin equations (3.49). To this end the Langevin equations are integrated formally over the small time interval $[t, t + \tau]$ after inserting the Taylor expansions:

$$k_i(\boldsymbol{x}(t + \tau')) = \sum_{s_1 \ldots s_c = 0}^{\infty} \frac{\Delta_1^{s_1}(\tau') \ldots \Delta_c^{s_c}(\tau')}{s_1! \ldots s_c!} \partial_1^{s_1} \ldots \partial_c^{s_c} k_i(\boldsymbol{x}_0(t)) \tag{3.64}$$

and

$$g_{ij}(\boldsymbol{x}(t + \tau')) = \sum_{s_1 \ldots s_c = 0}^{\infty} \frac{\Delta_1^{s_1}(\tau') \ldots \Delta_c^{s_c}(\tau')}{s_1! \ldots s_c!} \partial_1^{s_1} \ldots \partial_c^{s_c} g_{ij}(\boldsymbol{x}_0(t)) . \tag{3.65}$$

From (3.49) follows:

$$\Delta_i(\tau) = \int\limits_t^{t+\tau} \frac{dx_i(t+\tau')}{d\tau'}\,d\tau'$$

$$= \sum_{s_1 \ldots s_c = 0}^{\infty} \frac{\partial_1^{s_1} \ldots \partial_c^{s_c} k_i(\boldsymbol{x}_0(t))}{s_1! \ldots s_c!} \int\limits_t^{t+\tau} \Delta_1^{s_1}(\tau') \ldots \Delta_c^{s_c}(\tau')\,d\tau'$$

$$+ \sum_{j=1}^{c} \sum_{s_1 \ldots s_c = 0}^{\infty} \frac{\partial_1^{s_1} \ldots \partial_c^{s_c} g_{ij}(\boldsymbol{x}_0(t))}{s_1! \ldots s_c!} \int\limits_t^{t+\tau} \Delta_1^{s_1}(\tau') \ldots \Delta_c^{s_c}(\tau')\,\xi_j(t+\tau')\,d\tau' \,. \tag{3.66}$$

The equation (3.66) may be used as an iteration algorithm for determining $\Delta_i(\tau)$ to successive orders of accuracy: The $(n+1)$ th approximant $\Delta_{(n+1)i}(\tau)$ is obtained from (3.66) by inserting the nth approximant $\Delta_{(n)j}(\tau')$ on the rhs. Starting from the zeroth order $\Delta_{(0)i}(\tau) = 0$ the first and second order approximants obtained in this way are

$$\Delta_{(1)i}(\tau) = k_i(\boldsymbol{x}_0(t)) \int\limits_t^{t+\tau} d\tau' + \sum_{j=1}^{c} g_{ij}(\boldsymbol{x}_0(t)) \int\limits_t^{t+\tau} \xi_j(t+\tau')\,d\tau' \tag{3.67}$$

and

$$\Delta_{(2)i}(\tau) = \sum_{s_1 \ldots s_c = 0}^{\infty} \frac{\partial_1^{s_1} \ldots \partial_c^{s_c} k_i(\boldsymbol{x}_0(t))}{s_1! \ldots s_c!} \int\limits_t^{t+\tau} \Delta_{(1)1}^{s_1}(\tau') \ldots \Delta_{(1)c}^{s_c}(\tau')\,d\tau'$$

$$+ \sum_{j=1}^{c} \sum_{s_1 \ldots s_c = 0}^{\infty} \frac{\partial_1^{s_1} \ldots \partial_c^{s_c} g_{ij}(\boldsymbol{x}_0(t))}{s_1! \ldots s_c!} \int\limits_t^{t+\tau} \Delta_{(1)1}^{s_1}(\tau'') \ldots \Delta_{(1)c}^{s_c}(\tau'')\,\xi_j(t+\tau'')\,d\tau'' \,. \tag{3.68}$$

Finally, the moments follow by inserting $\Delta_i(\tau)$ (determined, say, to the nth order) into the definition (3.62) and by taking the limit $\tau \to 0$. The somewhat tedious but straight forward inspection of the moments $\langle \Delta_1^{s_1}(\tau) \ldots \Delta_c^{s_c}(\tau) \rangle$ leads to the surprising and highly welcome result that only the first and second moments $\langle \Delta_i(\tau) \rangle$ and $\langle \Delta_i(\tau)\,\Delta_j(\tau) \rangle$ contain terms proportional to τ, while the higher moments have only contributions of higher order in τ vanishing in the limit as $\tau \to 0$. Furthermore, the relevant terms $\sim\tau$ of $\langle \Delta_i(\tau) \rangle$ and $\langle \Delta_i(\tau)\,\Delta_j(\tau) \rangle$ can already be obtained by using the second order approximants (3.68). Making use of (3.61, 62, 68), these relevant terms are

$$\langle \Delta_i(\tau) \rangle = k_i(\boldsymbol{x}_0(t)) \int_t^{t+\tau} d\tau'$$

$$+ \sum_{j=1}^{c} \sum_{k=1}^{c} \partial_k g_{ij}(\boldsymbol{x}_0(t)) \sum_{l=1}^{c} g_{kl}(\boldsymbol{x}_0(t)) \int_t^{t+\tau} d\tau'' \int_t^{t+\tau''} d\tau' \langle \xi_l(t+\tau') \xi_j(t+\tau'') \rangle$$

$$= \tau k_i(\boldsymbol{x}_0) + \frac{\tau}{2} \sum_{j=1}^{c} \sum_{k=1}^{c} [\partial_k g_{ij}(\boldsymbol{x}_0)] g_{kj}(\boldsymbol{x}_0) \tag{3.69}$$

and

$$\langle \Delta_i(\tau) \Delta_k(\tau) \rangle$$

$$= \sum_{j=1}^{c} \sum_{l=1}^{c} g_{ij}(\boldsymbol{x}_0(t)) g_{kl}(\boldsymbol{x}_0(t)) \int_t^{t+\tau} d\tau' \int_t^{t+\tau} d\tau'' \langle \xi_j(t+\tau') \xi_l(t+\tau'') \rangle$$

$$= \tau \sum_{j=1}^{c} g_{ij}(\boldsymbol{x}_0) g_{kj}(\boldsymbol{x}_0) . \tag{3.70}$$

Eqs. (3.69, 70) yield the drift and fluctuation coefficients by making use of their definitions (3.30, 31)

$$K_i(\boldsymbol{x}) = k_i(\boldsymbol{x}) + \frac{1}{2} \sum_{k,j=1}^{c} [\partial_k g_{ij}(\boldsymbol{x})] g_{kj}(\boldsymbol{x}) \tag{3.71}$$

and

$$\varepsilon Q_{ij}(\boldsymbol{x}) = \sum_{j=1}^{c} g_{ij}(\boldsymbol{x}) g_{kj}(\boldsymbol{x}) . \tag{3.72}$$

Furthermore, the higher moments of the stochastic equation vanish:

$$\varepsilon^{s-1} M_{s_1 \ldots s_c}(\boldsymbol{x}) = 0 \qquad \text{for} \quad s = s_1 + \ldots + s_c \geq 3 . \tag{3.73}$$

The result expressed in (3.71–73) implies that the Langevin equations (3.49) with Gaussian δ-correlated random forces leads exactly to the Fokker-Planck equation (3.32).

Vice versa, if $g_{ij}(\boldsymbol{x})$ is presumed to be a positive definite matrix, (3.71, 72) may be inverted and $g_{ij}(\boldsymbol{x})$ and $k_i(\boldsymbol{x})$ determined from $\varepsilon Q_{ij}(\boldsymbol{x})$ and $K_i(\boldsymbol{x})$.

In this sense, the Fokker-Planck equation (3.32) and the Langevin equations (3.49) are equivalent descriptions. This equivalence was first derived by *Stratonovich*[3, 5]. For further reference see [1.1].

Although "Gaussian δ-correlated noise" is a reasonable standard assumption for many systems with fluctuations, it may be that sociology provides examples where the macrovariables are influenced by fluctuating forces of non-Gaussian nature and more complicated time correlation. In these cases the Langevin equations (3.49) are still an appropriate description, while the probability distribution $P(\boldsymbol{x}; t)$ obeys a more complicated equation than (3.32), for instance, the master equation or the general stochastic equation (3.29).

3.2.4 Approximate Mean Value Equations

The equations discussed in the preceding sections of this chapter fully describe – in terms of random forces or probability distributions – the effect of the unpredictability of single individual decisions or actions, even if assumptions about trends are made. But it is difficult to find their solutions, especially in multi-dimensional cases. Often it is simpler and sufficient to set up and solve equations for the *mean values* and eventually for the *variances*.

If the fundamental equation is a Fokker-Planck equation (3.32) the objective is to derive a closed set of equations for the mean values

$$\langle x_k \rangle_t = \int_0^\infty x_k P(\boldsymbol{x}; t)\, d^c\boldsymbol{x} \tag{3.74}$$

and the variances

$$\begin{aligned} \sigma_{lk} = \sigma_{kl} &= \langle (x_l - \langle x_l \rangle_t)(x_k - \langle x_k \rangle_t) \rangle_t \\ &= \langle x_l x_k \rangle_t - \langle x_l \rangle_t \langle x_k \rangle_t \end{aligned}$$

where

$$\langle x_l x_k \rangle_t = \int_0^\infty x_l x_k P(\boldsymbol{x}; t)\, d^c\boldsymbol{x} \tag{3.75}$$

The procedure adopted generalizes that of Chap. 2.

In the following it will be assumed that $P(\boldsymbol{x}; t)$ and its derivatives $\partial_i P(\boldsymbol{x}; t)$ are negligibly small at the boundary of the domain of definition of $P(\boldsymbol{x}; t)$; that is, where at least one component of $\boldsymbol{x}$, say x_i, is such that $x_i = 0$ or $x_i \to \infty$. Then the boundary terms in partial integrations can be neglected. This assumption is justified if the distribution $P(\boldsymbol{x}; t)$ centers about values $x_j = n_j/N_0$ with $n_j \gg 1$ for all $j = 1, 2, \ldots, C$ and vanishes for $x_j \to \infty$ because of the normalization. Further, the *sum convention* now used automatically takes the sum from 1 to C over an index appearing twice in an expression.

Using (3.32, 33) the equation of motion for $\langle x_k \rangle_t$ is obtained after partial integration as:

$$\begin{aligned} \frac{d\langle x_k \rangle_t}{dt} &= \int_0^\infty x_k \frac{\partial P(\boldsymbol{x}; t)}{\partial t}\, d^c\boldsymbol{x} = -\int_0^\infty x_k \partial_i I_i(\boldsymbol{x}; t)\, d^c\boldsymbol{x} \\ &= \int_0^\infty I_k(\boldsymbol{x}; t)\, d^c\boldsymbol{x} = \int_0^\infty K_k(\boldsymbol{x}) P(\boldsymbol{x}; t)\, d^c\boldsymbol{x} = \langle K_k(\boldsymbol{x}) \rangle_t\,. \end{aligned} \tag{3.76}$$

For a uni-modal distribution centered around $\langle \boldsymbol{x} \rangle_t$ it follows from the exact equation (3.76) that the approximate closed equation is:

$$\frac{d\langle x_k \rangle_t}{dt} \approx K_k(\langle \boldsymbol{x} \rangle_t) . \tag{3.77}$$

Proceeding in an analogous way

$$\begin{aligned}\frac{d\langle x_l x_k \rangle_t}{dt} &= -\int_0^\infty x_l x_k \partial_i I_i(\boldsymbol{x}; t)\, d^c\boldsymbol{x} \\ &= \int_0^\infty [x_k I_l(\boldsymbol{x}; t) + x_l I_k(\boldsymbol{x}; t)]\, d^c\boldsymbol{x} \\ &= \langle x_k K_l(\boldsymbol{x}) \rangle_t + \langle x_l K_k(\boldsymbol{x}) \rangle_t + \varepsilon \langle Q_{lk}(\boldsymbol{x}) \rangle_t \end{aligned} \tag{3.78}$$

Expanding $K_l(\boldsymbol{x})$ and $K_k(\boldsymbol{x})$ around $\langle \boldsymbol{x} \rangle_t$ in (3.78)

$$\begin{aligned}\langle x_k K_l(\boldsymbol{x}) \rangle_t &= \langle x_k \rangle_t K_l(\langle \boldsymbol{x} \rangle_t) + \langle x_k (x_i - \langle x_i \rangle_t) \rangle_t \partial_i K_l(\langle \boldsymbol{x} \rangle_t) \\ &= \langle x_k \rangle_t K_l(\langle \boldsymbol{x} \rangle_t) + \sigma_{ki}(t)\, \partial_i K_l(\langle \boldsymbol{x} \rangle_t) \end{aligned} \tag{3.79}$$

and using $\langle Q_{lk}(\boldsymbol{x}) \rangle_t \approx Q_{lk}(\langle \boldsymbol{x} \rangle_t)$ as well as (3.77), an approximate equation for the motion of $\sigma_{lk}(t)$ is obtained which is valid for an uni-modal distribution as

$$\begin{aligned}\frac{d\sigma_{lk}(t)}{dt} &\approx \varepsilon Q_{lk}(\langle \boldsymbol{x} \rangle_t) \\ &\quad + \sigma_{ik}(t)\, \partial_i K_l(\langle \boldsymbol{x} \rangle_t) + \sigma_{li}(t)\, \partial_i K_k(\langle \boldsymbol{x} \rangle_t) . \end{aligned} \tag{3.80}$$

If the $\langle x_k \rangle_t$ are known by solving (3.77), Eq. (3.80) is a system of non-homogeneous linear differential equations with given time dependent coefficients for the variance matrix $\sigma_{lk}(t)$.

The approximate equations (3.77, 80) account for the paths of the mean values and dispersion of distributions as long as the latter do not divide out into multi-modal distributions.

3.2.5 *Exact Mean Value Equations

Closed exact equations for the mean values $\langle \boldsymbol{x} \rangle_t$, whose probability distribution is $P(\boldsymbol{x}; t)$, will now be derived. As they remain valid *in all cases*, however, they have a more complicated structure than (3.77) which makes their evaluation more difficult.

---

* This section can be omitted on a first reading.

Firstly the Fokker-Planck equation and its formal solution is written in the more abstract form:

$$\partial_t P(\boldsymbol{x}; t) = \mathcal{L} P(\boldsymbol{x}; t) \qquad \text{with} \quad \partial_t \equiv \partial/\partial t \tag{3.81}$$

with the formal solution

$$P(\boldsymbol{x}; t) = \exp(\mathcal{L} t)\, P(\boldsymbol{x}; 0) \tag{3.82}$$

and the mean values and their time derivatives in the form of a scalar product

$$\langle f(\boldsymbol{x})\rangle_t = \int_0^\infty d^c\boldsymbol{x}\, P(\boldsymbol{x}; t) f(\boldsymbol{x}) \equiv (P|f) \tag{3.83}$$

$$\partial_t \langle f(\boldsymbol{x})\rangle_t = \int_0^\infty d^c\boldsymbol{x}\, \partial_t P(\boldsymbol{x}; t) f(\boldsymbol{x})$$

$$= (\mathcal{L}P|f) = (P|\mathcal{L}^+ f) \tag{3.84}$$

where $\mathcal{L}$ is the Fokker-Planck operator written with the summation convention as

$$\mathcal{L} F(\boldsymbol{x}) = [-\partial_i K_i(\boldsymbol{x}) + \frac{\varepsilon}{2}\partial_i \partial_j Q_{ij}(\boldsymbol{x})] F(\boldsymbol{x})$$

$$= \partial_i \left[\varepsilon L_{ij}(\boldsymbol{x})\partial_j - \varepsilon \left(\frac{\partial_j W(\boldsymbol{x})}{W(\boldsymbol{x})}\right) L_{ij}(\boldsymbol{x})\right] F(\boldsymbol{x}) \tag{3.85}$$

and $\mathcal{L}^+$ its adjoint[1], i.e.

$$\mathcal{L}^+ F(\boldsymbol{x}) = \left[K_i(\boldsymbol{x})\partial_i + \frac{\varepsilon}{2} Q_{ij}(\boldsymbol{x})\partial_i \partial_j\right] F(\boldsymbol{x})$$

$$= \varepsilon\left[\partial_j L_{ij}(\boldsymbol{x}) + L_{ij}(\boldsymbol{x})\left(\frac{\partial_j W(\boldsymbol{x})}{W(\boldsymbol{x})}\right)\right] \partial_i F(\boldsymbol{x})\,. \tag{3.86}$$

Secondly, a so-called "relevant" distribution $\overline{P}(\boldsymbol{x}; \boldsymbol{\lambda}(t))$ depending on time t via the C parameters $\boldsymbol{\lambda}(t) = \{\lambda_1(t), \ldots, \lambda_\psi(t)\}$ only is chosen. The relevant distribution does not in general agree with the (unknown) exact solution of (3.81) but has to satisfy the conditions of being properly normalized

$$\int_0^\infty d^c\boldsymbol{x}\, \overline{P}(\boldsymbol{x}; \boldsymbol{\lambda}(t)) \equiv (\overline{P}|1) = 1 \tag{3.87}$$

[1] Again boundary terms in partial integrations are neglected.

and of agreeing with the exact distribution $P(\boldsymbol{x}; t)$ at the time $t = 0$:

$$\bar{P}(\boldsymbol{x}; \boldsymbol{\lambda}(0)) = P(\boldsymbol{x}; 0) \tag{3.88}$$

and lastly, but most importantly, of being *macro-equivalent* with $P(\boldsymbol{x}; t)$ for all times t in the sense that mean values of $x_1, \ldots, x_c$ taken over $\bar{P}$ agree with the *exact* mean values taken over P:

$$\begin{aligned} \langle x_j \rangle_\lambda &= (\bar{P}(t)|x_j) \equiv \int_0^\infty d^c\boldsymbol{x}\, \bar{P}(\boldsymbol{x}; \boldsymbol{\lambda}(t))\, x_j \\ &= \langle x_j \rangle_t = (P(t)|x_j) \equiv \int_0^\infty d^c\boldsymbol{x}\, P(\boldsymbol{x}; t)\, x_j \\ &\qquad \text{for} \quad j = 1, 2, \ldots, C\,. \end{aligned} \tag{3.89}$$

Equation (3.89) expresses the $\langle x_j \rangle_t$ by means of the parameters $\{\lambda_1(t), \ldots, \lambda_c(t)\}$ and should be capable of being inverted in order to express the $\lambda_i(t)$ in terms of the mean values $\{\langle x_1 \rangle_t \ldots \langle x_c \rangle_t\}$. Hence, $\bar{P}(\boldsymbol{x}, \boldsymbol{\lambda}(t))$ can also be considered as a function of the mean values $\langle x_j \rangle_t$.

Thirdly, a closed exact set of equations has to be set up for the – as yet – unknown mean values or, equivalently, for the parameters $\{\lambda_1(t), \ldots, \lambda_c(t)\}$. This can be achieved by evaluating the expression

$$J_j(t) = \int_0^t \partial_s (\exp(\mathcal{L}s)\, \bar{P}(t-s)|\mathcal{L}^+ x_j)\, ds \tag{3.90}$$

in two different ways. Firstly integrating the rhs of (3.90) from $s = 0$ to $s = t$ leads to

$$\begin{aligned} J_j(t) &= (\exp(\mathcal{L}s)\, \bar{P}(t-s)|\mathcal{L}^+ x_j)|_{s=0}^{s=t} \\ &= (\mathcal{L} \exp(\mathcal{L}t)\, P(0)|x_j) - (\bar{P}(t)|\mathcal{L}^+ x_j) \\ &= \partial_t \langle x_j \rangle_t - \langle K_j(\boldsymbol{x}) \rangle_\lambda \end{aligned} \tag{3.91 a}$$

where (3.81, 82) and $\mathcal{L}^+ x_j = K_j(\boldsymbol{x})$ have been used. Secondly performing the differentiation ∂_s within the integrand of (3.90) yields

$$J_j(t) = \int_0^t ([\mathcal{L}\bar{P}(t-s) + \partial_s \bar{P}(t-s)]|\exp(\mathcal{L}^+ s)\, K_j)\, ds \tag{3.91 b}$$

Equating (3.91 a, b) the result

$$\begin{aligned} \partial_t \langle x_j \rangle_t &= \langle K_j(\boldsymbol{x}) \rangle_\lambda \\ &\quad + \int_0^t ([\mathcal{L}\bar{P}(t-s) + \partial_s \bar{P}(t-s)]|\exp(\mathcal{L}^+ s)\, K_j)\, ds \end{aligned} \tag{3.92}$$

is obtained. Comparing (3.92) with (3.76)

$$\partial_t \langle \boldsymbol{x} \rangle_t = \langle K_j(\boldsymbol{x}) \rangle_t$$

which are both exact equations, it can be seen that the second term of the integro-differential equation (3.92) accounts for the difference between $\langle K_j(\boldsymbol{x}) \rangle_t$ and $\langle K_j(\boldsymbol{x}) \rangle_\lambda$, i.e. for the difference between the exact and the "relevant" mean value of the drift $K_j(\boldsymbol{x})$.

The decisive difference between (3.76) and (3.92) is that the latter represents a *closed* (non-linear integro-differential) set of equations for the $\langle x_j \rangle_t$ since according to the definition of $\overline{P}(\boldsymbol{x}, \boldsymbol{\lambda}(t))$ the rhs of (3.92) only depends on $\langle \boldsymbol{x} \rangle_{t-s}$ for $0 \leq s \leq t$.

It turns out that an appropriate choice for the relevant distribution is

$$\overline{P}(\boldsymbol{x}; \boldsymbol{\lambda}(t)) = Z^{-1}(\boldsymbol{\lambda}(t))\, W(\boldsymbol{x}) \exp(-\lambda_i(t)\, x_i) \tag{3.93}$$

where $W(\boldsymbol{x})$ is the stationary solution to (3.34) and $Z(\boldsymbol{\lambda})$ is the normalization factor. For this choice it follows using (3.85) that:

$$\mathscr{L}\overline{P}(\boldsymbol{x}; \boldsymbol{\lambda}(t)) = -\lambda_j(t)\, \partial_i [\varepsilon L_{ij}(\boldsymbol{x}) \overline{P}(\boldsymbol{x}; \boldsymbol{\lambda}(t))] \tag{3.94}$$

and

$$\partial_t \overline{P}(\boldsymbol{x}; \boldsymbol{\lambda}(t)) = -[\partial_t \lambda_i(t)](x_i - \langle x_i \rangle_\lambda)\, \overline{P}(\boldsymbol{x}; \boldsymbol{\lambda}(t)) \,. \tag{3.95}$$

Insertion of (3.94, 95) in the rhs of (3.92) leads to the more explicit form of the exact closed equations of motion for the mean values, viz:

$$\begin{aligned} \partial_t \langle x_k \rangle_t = & \int_0^\infty d^c\boldsymbol{x}\, \overline{P}(\boldsymbol{x}; \boldsymbol{\lambda}(t))\, K_k(\boldsymbol{x}) \\ & - \int_0^t ds \int_0^\infty d^c\boldsymbol{x}\, \{[\partial_i \varepsilon L_{ij}(\boldsymbol{x}) \overline{P}(\boldsymbol{x}; \boldsymbol{\lambda}(t-s))]\, \lambda_j(t-s) \\ & + (\partial_s \lambda_i(t-s))(x_i - \langle x_i \rangle_\lambda)\, \overline{P}(\boldsymbol{x}; \boldsymbol{\lambda}(t-s))\} \exp(\mathscr{L}^+ s)\, K_k(\boldsymbol{x}) \,. \end{aligned} \tag{3.96}$$

Equations of the type (3.96) have been recently set up for general Fokker-Planck systems and evaluated for special models [3.6, 7].

3.3 The Dynamics of Trend Parameters and of the Situation Vector

The equations of motion for the socio-configuration do not yet reflect the full complexity of the dynamics of a socio-system. Nevertheless they describe the motion of collective psychological and political processes (namely the changes of attitudes) under the influence of a given situation vector $\boldsymbol{y}$ and a given set of trend parameters $\boldsymbol{\kappa}$. In Sect. 3.2 these equations have been used under the assumption that $\boldsymbol{y}$ and $\boldsymbol{\kappa}$ are constants in time or at most parameters of a given slow time dependence. Correspondingly, the transition probabilities (3.12) or the drift and fluctuation coefficients (3.30, 31) were either constants in time or only loosely time dependent via the parameters $\boldsymbol{y}$ and $\boldsymbol{\kappa}$.

If they are constants with time, it can be shown (e.g. by using Boltzmann's famous H-theorem) that all distributions developing according to the master equation (3.14) or the Fokker-Planck equation (3.32) or (3.34) end up in *the stationary distribution* $W(\boldsymbol{x})$. After this the motion of a sample socio-configuration can only consist of stationary fluctuations around the probability peaks of the stationary distribution. While this kind of motion may describe a "static society" under stable psychological and situational conditions, it does not comprise the full scope of sociological phenomena.

The more general case, as already indicated in Sect. 3.1, consists of an interaction between socio-configuration $\boldsymbol{x}(t) \in \mathfrak{C}$, situation vector $\boldsymbol{y}(t) \in \mathfrak{S}$ and the trend parameters $\boldsymbol{\kappa}(t) = \{\kappa_1(t), \ldots, \kappa_T(t)\} \in \mathfrak{T}$ where the trend parameters $\boldsymbol{\kappa}(t)$ are now represented as elements of a T-dimensional *trend-space* $\mathfrak{T}$. When thinking in terms of Langevin equations the coupled set of equations for the variables

$$\{\boldsymbol{x}(t), \boldsymbol{y}(t), \boldsymbol{\kappa}(t)\} \in \mathfrak{C} \oplus \mathfrak{S} \oplus \mathfrak{T} \equiv \mathfrak{D} \tag{3.97}$$

has to be considered. The total space $\mathfrak{D}$ of the dynamic variables now includes not only the variables $\boldsymbol{x}(t)$ of the socio-configuration treated in Sect. 3.2 but also the variables $\boldsymbol{y}(t)$ of the material situation and $\boldsymbol{\kappa}(t)$ of the psychological trends.

Here only some general remarks are possible on the structure of the equations of motion for the variables $\boldsymbol{y}(t)$ and $\boldsymbol{\kappa}(t)$ whose concrete form has to be specifically worked out for the different separable sectors of society.

The situation variable $\boldsymbol{y}(t)$ will be "driven" by the activities of people of differing influence and differing attitudes. Therefore the time derivative of $\boldsymbol{y}(t)$ will be a superposition of partial rates due to these competing influences:

$$\frac{dy_i}{dt} = \left(\frac{\partial y_i}{\partial t}\right)_1 + \ldots + \left(\frac{\partial y_i}{\partial t}\right)_\Omega \qquad (i = 1, 2, \ldots, S) \,. \tag{3.98}$$

Each term describes a certain partial influence and will in general depend, non-linearly on $\boldsymbol{y}(t)$, the socio-configuration $\boldsymbol{x}(t)$ and certain control parameters whose meaning is implicitly defined by (3.98).

Similarly, the time derivatives of the trend parameters $\boldsymbol{\kappa}(t)$ (which determine the explicit form of the transition probabilities) describe the psychological and political reactions of trends to the changing situation and will also consist of a superposition of competing influences expressed in partial rates:

$$\frac{d\kappa_l}{dt} = \left(\frac{\partial \kappa_l}{\partial t}\right)_1 + \ldots + \left(\frac{\partial \kappa_l}{\partial t}\right)_\Gamma \qquad (l = 1, 2, \ldots, T) \,. \tag{3.99}$$

For example, a partial rate can describe an opposing change in trend provoked by the situation $\boldsymbol{y}(t)$. This altered trend will then lead to an alteration in the socio-configuration which in turn drives the situation $\boldsymbol{y}(t)$ into a new position generating the next reaction, and so on. So far, terms with random forces have been neglected in (3.98, 99) which would transform these equations into Langevin equations and lead to a statistical description analogous to that for $\boldsymbol{x}(t)$ in Sect. 3.2. It is suggested that a multi-dimensional periodic or quasi-periodic or even chaotic motion for $\boldsymbol{x}(t) \oplus \boldsymbol{y}(t) \oplus \boldsymbol{\kappa}(t) \in \mathfrak{D}$ is a reasonable and justifiable type of solution for the coupled equations of the socio-configuration, the material situation and the psychological trends for the medium term time development.

Indeed, there exist systems of nonlinear coupled differential equations with *"limit cycles"* or *"strange attractors"*, whose solutions exhibit just this sort of quasi-periodic or chaotic behaviour. Simple examples of such equations will be found in the following chapters.

Thinking of the long-term time development of a situation, however, the superposition of a long-term irreversible motion can be expected to transform the quasi-periodic and quasi-closed path into a (multi-dimensional) non-closed semi-helical path winding through $\mathfrak{D}$. (The words "progress" or "regress" for this irreversible motion have been deliberately avoided as they imply the valuations of "improvement" or "deterioration" of a situation, whereas required is a description of the "dynamics as such" before making valuations.) The irreversible motion for instance may be due to the evolution of the genetic constitution, to the evolution of a "cultural memory" including the development of technologies or to the exhaustion of natural resources, etc.

Finally, the problem of the *"hierarchy of trend parameters* $\boldsymbol{\kappa}$*"* will be discussed. The problem is common to all theories starting from a phenomenological level and consists of the following: If dynamic equations are set up for the trend parameters by considering their coupling to the situation, new parameters have to be introduced which can be denoted as trend parameters of second order. If the latter are also considered as dynamic variables, in principle a "regressus ad infinitum" will be obtained: The dynamic equations for trend parameters of nth. order will contain trend parameters of $(n + 1)$th order for $n = 1, 2, 3, \ldots$. Of course, this potentially infinite series would have to be truncated at an appropriate nth. order or level where it is reasonable to treat the $(n + 1)$th trend parameters as phenomenologically given constants.

3.4 *Mean Value Equations for Grossvariables of the Socio-Configuration

Before starting with the topic of this section, the derivation of mean value equations for the socio-configuration $\{n_{ai}\}$ has to be reviewed.

Starting from the master equation (3.14) and after the introduction of scaled variables (3.25) the stochastic equation (3.29) and the Fokker-Planck equation

* This section can be omitted on a first reading.

(3.32) were derived. From the latter exact non-closed equations (3.76) and finally approximate closed equations (3.77) for the mean values $\tilde{x}_{\alpha i} \equiv \langle x_{\alpha i} \rangle_t$ or $\tilde{n}_{\alpha i} \equiv \langle n_{\alpha i} \rangle_t$ of the socio-configuration variables were established. These equations become explicit if the definition (3.30) for the drift coefficients $K_j(\boldsymbol{x})$ and the explicit form of the transition probabilities $w[\boldsymbol{k}, \boldsymbol{n}]$ composed of the partial contributions (3.16, 18, 20) is used.

Alternatively, the mean value equations for $\tilde{n}_{\alpha i}$ may be derived directly, i.e. without using the Fokker-Planck equation, from the master equation. This can be seen in Chap. 4 and leads from the master equation (4.6) directly to the approximate closed mean value equations (4.14). The result is equivalent to that derived via the Fokker-Planck equation.

Omitting birth/death processes but taking into account individual transitions between attitudes and between subpopulations as described by transition probabilities p_{ji}^{α} *and* $p_i^{\beta\alpha}$ see (3.15, 17), and by putting $g = 1$ for simplicity, the following approximate closed mean value equations in full correspondence to their explicit form (4.14) are obtained

$$\frac{d\tilde{n}_{\gamma k}}{dt} = \left\{ \sum_{i=1}^{L} \tilde{n}_{\gamma i} p_{ki}^{\gamma}(\tilde{\boldsymbol{n}}, \boldsymbol{y}, \boldsymbol{\kappa}) - \sum_{j=1}^{L} \tilde{n}_{\gamma k} p_{jk}^{\gamma}(\tilde{\boldsymbol{n}}, \boldsymbol{y}, \boldsymbol{\kappa}) \right\}$$

$$+ \left\{ \sum_{\alpha=1}^{P} \tilde{n}_{\alpha k} p_k^{\gamma\alpha}(\tilde{\boldsymbol{n}}, \boldsymbol{y}, \boldsymbol{\kappa}) - \sum_{\alpha=1}^{P} \tilde{n}_{\gamma k} p_k^{\alpha\gamma}(\tilde{\boldsymbol{n}}, \boldsymbol{y}, \boldsymbol{\kappa}) \right\} . \tag{3.100}$$

Here, the transition probabilities in general can not only depend on the socio-configuration $\tilde{\boldsymbol{n}}$ and on trend parameters $\boldsymbol{\kappa}$ but also on the material situation $\boldsymbol{y}$. The first term of the rhs of (3.100) conserves the number

$$N_\alpha = \sum_{i=1}^{L} n_{\alpha i}, \quad \alpha = 1, 2, \ldots, P$$

and the second term conserves the number

$$N_i = \sum_{\alpha=1}^{P} n_{\alpha i}, \quad i = 1, 2, \ldots, L \ .$$

If, however, the $\tilde{n}_{\gamma k}$ change under the influence of both terms, only the total number

$$N = \sum_{\alpha=1}^{P} \sum_{i=1}^{L} n_{\alpha i}$$

remains constant with time.

The problem to be stated is: Assuming L attitudes and P subpopulations, where $L \gg 1$ and $P \geq 1$, the description of the society using the $C = PL$ mean values $\{\tilde{n}_{\alpha i}\}$ of the socio-configuration variables $\{n_{\alpha i}\}$ is still a rather *fine grained* and hence complicated one. Therefore it is the intention to derive from (3.100) even more approximate, but closed, equations for the mean values $\tilde{G}_q(t)$ of *a few grossvariables of the attitude situation* $G_q(t)$, $(q = 1, 2, \ldots, Q)$ which are as "coarse grained" as the macrovariables $y_i(t)$, $(i = 1, 2, \ldots, S)$ of the *material situation*.

These grossvariables should be functions of the C socio-configuration variables $\{n_{\alpha i}\}$. Although their number should be much smaller than that of the $\{n_{\alpha i}\}$ – i.e. $Q \ll C$ – they should be carefully chosen in order to characterize the global state of the society for the aspect under consideration.

The following general form of the grossvariables is assumed:

$$G_q = \sum_{i=1}^{L} \sum_{\alpha=1}^{P} g_{\alpha i}^q n_{\alpha i} \tag{3.101}$$

where $g_{\alpha i}^q$ are real coefficients for $q = 1, 2, \ldots, Q$. This implies that the G_q are appropriately chosen *weighted averages* over the numbers of people of subpopulations having certain attitudes. The numbers $N_\alpha(\alpha = 1, 2, \ldots, P)$ and $N_i(i = 1, 2, \ldots, L)$ are special cases of such grossvariables. An example illustrates the implications of (3.101): In Chap. 6 the grossvariables P_1, P_2 will be introduced which characterize the global state of "political psychology" ranging from "sympathy" to "tension" between two interacting competitive macro-societies $\mathcal{S}_j$. If P_j is to have the meaning that positive (negative) values describe the prevalence of sympathy (tension) in $\mathcal{S}_j$, it is clear how the weights in the definition of P_j have to be chosen. For instance, positive weights $g_{\alpha i}^j > 0$ have to be allocated to attitude levels (αi) such as "cooperativeness" and "conciliativeness" and negative weights $g_{\beta k}^j < 0$ to attitude levels (βk) such as "militance" and "ideological dogmatism".

Coming back to the stated problem, it is possible to derive – as a first step – *non-closed* equations of motion for the mean values $\tilde{G}_q(t)$. Multiplying (3.100) with $g_{\gamma k}^q$ and taking the sum over γ and k

$$\begin{aligned}\frac{d}{dt}\sum_{\gamma,k}^{P,L} g_{\gamma k}^q \tilde{n}_{\gamma k} &\equiv \frac{d}{dt}\tilde{G}_q \\ &= \sum_{\gamma,k}^{P,L} g_{\gamma k}^q \left[\sum_{i=1}^{L} \tilde{n}_{\gamma i} p_{ki}^{\gamma}(\tilde{\boldsymbol{n}}, \boldsymbol{y}, \boldsymbol{\kappa}) - \sum_{j=1}^{L} \tilde{n}_{\gamma k} p_{jk}^{\gamma}(\tilde{\boldsymbol{n}}, \boldsymbol{y}, \boldsymbol{\kappa})\right] \\ &+ \sum_{\gamma,k}^{P,L} g_{\gamma k}^q \left[\sum_{\alpha=1}^{P} \tilde{n}_{\alpha k} p_k^{\gamma\alpha}(\tilde{\boldsymbol{n}}, \boldsymbol{y}, \boldsymbol{\kappa}) - \sum_{\alpha=1}^{P} \tilde{n}_{\gamma k} p_k^{\alpha\gamma}(\tilde{\boldsymbol{n}}, \boldsymbol{y}, \boldsymbol{\kappa})\right]\end{aligned} \tag{3.102}$$

is obtained. The rhs of (3.102) depends on all $\tilde{n}_{\gamma i}$ and not only on the $\tilde{G}_q$; hence (3.102) is non-closed.

In order to arrive at *closed* equations for the $\tilde{G}_q$ further assumptions and approximations have to be made.

At first an (almost) general form for the individual transition probabilities p^{γ}_{ki} and $p^{\gamma a}_{i}$ will be assumed consisting of positive definite exponential expressions whose exponents are Taylor expansions in the variables $\boldsymbol{n}$ and $\boldsymbol{y}$ of which only the linear terms will be considered:[2]

$$p^{\gamma}_{ki}(\boldsymbol{n}, \boldsymbol{y}, \boldsymbol{\kappa}) = \nu \exp\left[\delta\binom{\gamma}{ki} + \sum_{\beta,j}^{P,L} \kappa\left(\binom{\gamma}{ki}\middle|\beta j\right) n_{\beta j} + \sum_{m}^{S} \sigma\left(\binom{\gamma}{ki}\middle| m\right) y_m\right] \tag{3.103}$$

and

$$p^{\gamma a}_{i}(\boldsymbol{n}, \boldsymbol{\gamma}, \boldsymbol{\kappa}) = \mu \exp\left[\delta\binom{\gamma a}{i} + \sum_{\beta,j}^{P,L} \kappa\left(\binom{\gamma a}{i}\middle|\beta j\right) n_{\beta j} + \sum_{m}^{S} \sigma\left(\binom{\gamma a}{i}\middle| m\right) y_m\right] . \tag{3.104}$$

Here the coefficients ν, $\delta\binom{\gamma}{ki}$, $\kappa\left(\binom{\gamma}{ki}\middle|\beta j\right)$ and $\sigma\left(\binom{\gamma}{ki}\middle| m\right)$ as well as μ, $\delta\binom{\gamma a}{i}$, $\kappa\left(\binom{\gamma a}{i}\middle|\beta j\right)$ and $\sigma\left(\binom{\gamma a}{i}\middle| m\right)$ are trend parameters which will not be further specified at the moment.

If the grossvariables are expected to (approximately) obey a closed subdynamics, it is self-consistently plausible that the transition probabilities p^{γ}_{ki} and $p^{\gamma a}_{i}$ only depend on $\{n_{ai}\}$ via the gross variables G_q. This assumption amounts to a specialized choice of the trend parameters

$$\left.\begin{aligned} \kappa\left(\binom{\gamma}{ki}\middle|\beta j\right) &= \sum_{q=1}^{Q} \omega\left(\binom{\gamma}{ki}\middle| q\right) g^{q}_{\beta j} \\ &\text{and} \\ \kappa\left(\binom{\gamma a}{i}\middle|\beta j\right) &= \sum_{q=1}^{Q} \omega\left(\binom{\gamma a}{i}\middle| q\right) g^{q}_{\beta j} \end{aligned}\right\} \tag{3.105}$$

leading to

$$p^{\gamma}_{ki}(\boldsymbol{n}, \boldsymbol{y}, \boldsymbol{\kappa}) = p^{\gamma}_{ki}[\boldsymbol{G}, \boldsymbol{y}, \boldsymbol{\kappa}] = \nu \exp\left[\delta\binom{\gamma}{ki} + \sum_{q=1}^{Q} \omega\left(\binom{\gamma}{ki}\middle| q\right) G_q + \sum_{m=1}^{S} \sigma\left(\binom{\gamma}{ki}\middle| m\right) y_m\right] \tag{3.106}$$

---

[2] Expressions of this kind for the transition probabilities prove to be reasonable and sufficiently flexible even though they do not have a completely general form (Chaps. 2, 4, and 5)

and

$$p_i^{\gamma\alpha}(\boldsymbol{n}, \boldsymbol{y}, \boldsymbol{\kappa}) = p_i^{\gamma\alpha}[\boldsymbol{G}, \boldsymbol{y}, \boldsymbol{\kappa}]$$

$$= \mu \exp\left[\delta(i^{\gamma\alpha}) + \sum_{q=1}^{Q} \omega(i^{\gamma\alpha}|q) G_q + \sum_{m=1}^{S} \sigma(i^{\gamma\alpha}|m) y_m\right] . \quad (3.107)$$

Even this specialization for p_{ki}^{γ} and $p_i^{\gamma\alpha}$ does not yet lead to closed equations for $\tilde{G}_q$ as the rhs of (3.102) not only depends on $p_{ki}^{\gamma}[\tilde{\boldsymbol{G}}, \boldsymbol{y}, \boldsymbol{\kappa}]$ and on $p_i^{\gamma\alpha}[\tilde{\boldsymbol{G}}, \boldsymbol{y}, \boldsymbol{\kappa}]$ but also on each $\tilde{n}_{\alpha j}$ ($\alpha = 1, \ldots, P$; $j = 1, \ldots, L$) separately. Only if it can be justified to approximate the mean values $\tilde{n}_{\alpha i}(t)$ ($\alpha = 1, \ldots, P$; $j = 1, \ldots, L$) by functions of the mean values $\tilde{G}_q(t)$ only – i.e. by $\tilde{n}_{\alpha i}(t) = \tilde{n}_{\alpha i}[\tilde{\boldsymbol{G}}(t)]$ – can *approximate closed equations*:

$$\frac{d}{dt}\tilde{G}_q(t)$$

$$= \sum_{\gamma,k}^{P,L} g_{\gamma k}^{q}\left[\sum_{i=1}^{L} \tilde{n}_{\gamma i}[\tilde{\boldsymbol{G}}] p_{ki}^{\gamma}[\tilde{\boldsymbol{G}}, \boldsymbol{y}, \boldsymbol{\kappa}] - \sum_{j=1}^{L} \tilde{n}_{\gamma k}[\tilde{\boldsymbol{G}}] p_{jk}^{\gamma}[\tilde{\boldsymbol{G}}, \boldsymbol{y}, \boldsymbol{\kappa}]\right]$$

$$+ \sum_{\gamma,k}^{P,L} g_{\gamma k}^{q}\left[\sum_{\alpha=1}^{P} \tilde{n}_{\alpha k}[\tilde{\boldsymbol{G}}] p_k^{\gamma\alpha}[\tilde{\boldsymbol{G}}, \boldsymbol{y}, \boldsymbol{\kappa}] - \sum_{\alpha=1}^{P} \tilde{n}_{\gamma k}[\tilde{\boldsymbol{G}}] p_k^{\alpha\gamma}[\tilde{\boldsymbol{G}}, \boldsymbol{y}, \boldsymbol{\kappa}]\right] \quad (3.108)$$

for the reduced set of grossvariable mean values $\tilde{G}_q(t)$, ($q = 1, 2, \ldots, Q$) be obtained by inserting this assumption and (3.106, 107) into (3.102). In order to determine such functions $\tilde{n}_{\alpha i}[\tilde{\boldsymbol{G}}]$, methods are introduced which are fundamental in the statistical physics of equilibrium but which will now be applied mutatis mutandis to a *non-equilibrium situation* in the sociological framework being considered [3.8–11].

Evidently, only such values $n_{\alpha i}(t)$ of the socio-configuration are compatible with given values $G_q(t)$ – or better mean values $\tilde{G}_q(t)$ of the grossvariables at time t – which fulfill

$$\sum_{\alpha,i}^{P,L} g_{\alpha i}^{q} n_{\alpha i}(t) = \tilde{G}_q(t) \quad \text{for} \quad q = 1, 2, \ldots, Q . \quad (3.109)$$

Now each socio-configuration $\{n_{\alpha i}\}$ is already a *macro-configuration* which can be realized by many *micro-configurations*. Here by definition a *micro-configuration* is one special distribution of individuals into PL "boxes", where the box (α, i) represents subpopulation α and attitude i. A simple combinatorial consideration leads to the following formula for the *number* $M(\{n_{\alpha i}\})$ *of micro-configurations* belonging to the macro-configuration $\{n_{\alpha i}\}$ with $n_{\alpha i}$ individuals in "box" (α, i):

$$M(\{n_{\alpha i}\}) = \frac{N!}{\prod_{\alpha,i}^{P,L} n_{\alpha i}!} . \quad (3.110)$$

Furthermore it is reasonable to assume that the mean values $\tilde{G}_q(t)$ of the grossvariables have only slowly changing time paths as compared with the *rapid* transitions between different micro-configurations. Under this assumption it is plausible that the latter transitions establish a *momentary equilibrium* between all allowed micro-configurations for given values of $\tilde{G}_q(t)$. A second assumption is that the individual transition probabilities between all micro-configurations fulfilling the constraints (3.109) are of the same order of magnitude. Under this condition approximately the *same weight will be allocated* in the momentary equilibrium to *every micro-configuration* compatible with the momentary mean values $\tilde{G}_q(t)$ of the grossvariables. As a consequence, the probability of the macro-configuration $\{n_{\alpha i}\} = \boldsymbol{n}$ under the constraint of given momentary mean values $\tilde{G}_q(t)$ of grossvariables is then simply proportional to the number $M(\boldsymbol{n})$ of micro-configurations through which it can be realized. This probability is given by

$$\left.\begin{aligned} &p_G(\boldsymbol{n}) = \frac{M(\boldsymbol{n})}{\sum\limits_{\{\boldsymbol{n}\}}^{(G)} M(\boldsymbol{n})} \\ &\text{if } \boldsymbol{n} \equiv \{n_{\alpha i}\} \text{ is compatible with (3.109) and by} \\ &p_G(\boldsymbol{n}) = 0 \\ &\text{if } \boldsymbol{n} \equiv \{n_{\alpha i}\} \text{ is incompatible with (3.109).} \end{aligned}\right\} \tag{3.111}$$

Here the sum $\sum\limits_{\{\boldsymbol{n}\}}^{(G)}$ extends over all macro-configurations $\boldsymbol{n} = \{n_{\alpha i}\}$ compatible with the constraints (3.109).

Making use of (3.111) the mean values $\tilde{n}_{\alpha i}(t)$ under the constraints (3.109) by using their definition

$$\tilde{n}_{\beta k}(t) = \sum_{\{\boldsymbol{n}\}} n_{\beta k}\, p_{G(t)}(\boldsymbol{n}) \tag{3.112}$$

can – in principle – now be calculated. Since the probability distribution (3.111) depends on $\tilde{G}_q(t)$, the mean values $\tilde{n}_{\beta k}(t)$ of the socio-configuration now become functions of the mean values $\tilde{G}_q(t)$ of the grossvariables.

In practice it is cumbersome to evaluate the mean values (3.112) in this way. Fortunately, however, the probability distribution (3.111) has a definite or steep maximum for values $\{\hat{n}_{\alpha i}\}$ which maximize $M(\{n_{\alpha i}\})$, or equivalently $\ln[M(\{n_{\alpha i}\})]$ under the constraints (3.109). Because of the steepness of the maximum of $p_G(\{n_{\alpha i}\})$ the *mean values* $\tilde{n}_{\beta k}$ (approximately) agree with *the values* $\hat{n}_{\beta k}$ *maximizing* $p_G(\{n_{\alpha i}\})$.

The $\hat{n}_{\beta k}$ can be easily determined: Making use of Stirling's formula:

$$\ln(n!) \cong n\ln(n) - n \tag{3.113}$$

the following relation can be obtained

$$\ln\left(\frac{N!}{\prod_{\alpha,i}^{P,L} n_{\alpha i}!}\right) = [N\ln(N) - N] - \sum_{\alpha,i}^{P,L} [n_{\alpha i}\ln(n_{\alpha i}) - n_{\alpha i}] \,. \tag{3.114}$$

This expression has to be maximized under the condition:

$$\sum_{\alpha,i}^{P,L} n_{\alpha i} = N \tag{3.115}$$

and under the Q constraints (3.109). This leads to the variational principle

$$\left.\begin{aligned} &\delta\left\{-\sum_{\alpha,i}^{P,L}[n_{\alpha i}\ln(n_{\alpha i}) - n_{\alpha i}]\right.\\ &\quad\left. + \lambda_0\left(\sum_{\alpha,i}^{P,L} n_{\alpha i} - N\right) + \sum_{q=1}^{Q}\lambda_q\left[\sum_{\alpha,i}^{P,L} g_{\alpha i}^{q} n_{\alpha i} - \tilde{G}_q(t)\right]\right\} = 0 \end{aligned}\right\} \tag{3.116}$$

where λ_0 and λ_q, $(q = 1, 2, \ldots, Q)$ are Lagrangian parameters and where all $n_{\alpha i}$ can be varied independently. The variation leads to

$$\left.\begin{aligned} &\delta n_{\alpha i}\left[-\ln(n_{\alpha i}) + \lambda_0 + \sum_{q=1}^{Q}\lambda_q g_{\alpha i}^{q}\right] = 0 \qquad \text{for}\\ &n_{\alpha i} = \hat{n}_{\alpha i}, \qquad \text{or to} \quad \hat{n}_{\alpha i} = \exp\left(\lambda_0 + \sum_{q=1}^{Q}\lambda_q g_{\alpha i}^{q}\right) . \end{aligned}\right\} \tag{3.117}$$

Here the Lagrangian parameters λ_0 and λ_q are to be considered as functions of N and $\tilde{G}_q(t)$, $(q = 1, 2, \ldots, Q)$, i.e. $\lambda_0 = \lambda_0(N, \tilde{\boldsymbol{G}})$ and $\lambda_q = \lambda_q(N, \tilde{\boldsymbol{G}})$, which are determined by the constraints

$$\sum_{\alpha,i}^{P,L}\hat{n}_{\alpha i} \equiv \sum_{\alpha,i}^{P,L}\exp\left(\lambda_0 + \sum_{p=1}^{Q}\lambda_p g_{\alpha i}^{p}\right) = N \tag{3.118 a}$$

and

$$\sum_{\alpha,i}^{P,L} g_{\alpha i}^{q}\hat{n}_{\alpha i} \equiv \sum_{\alpha,i}^{P,L} g_{\alpha i}^{q}\exp\left(\lambda_0 + \sum_{p=1}^{Q}\lambda_p g_{\alpha i}^{p}\right) = \tilde{G}_q(t) \tag{3.118 b}$$

where $q = 1, 2, \ldots, Q$. Inserting the Lagrangian parameters $\lambda_0(N, \tilde{\boldsymbol{G}})$ and $\lambda_q(N, \tilde{\boldsymbol{G}})$ into (3.117) the most probable socio-configuration $\{\hat{n}_{\alpha i}\} \approx \{\tilde{n}_{\alpha i}\}$ is obtained as a function of N and $\tilde{G}_q$, $(q = 1, \ldots, Q)$. The functions $\hat{n}_{ai}(N, \tilde{G})$ can be inserted in (3.108) to yield the wanted set of equations of motion for $\tilde{G}_q(t)$.

In order to illustrate the analogy to thermodynamic relations the quantity

$$\hat{S}(t) = -\sum_{a,i}^{P,L} [\hat{n}_{ai}(t) \ln(\hat{n}_{ai}(t)) - \hat{n}_{ai}(t)] \tag{3.119}$$

can be introduced which bears some similarity to the entropy function in statistical physics and which by differentiating and inserting (3.117) gives

$$\begin{aligned} \frac{d\hat{S}(t)}{dt} &= -\sum_{a,i}^{P,L} \frac{d\hat{n}_{ai}}{dt} \ln(\hat{n}_{ai}) = -\sum_{a,i}^{P,L} \frac{d\hat{n}_{ai}}{dt}\left(\lambda_0 + \sum_{q=1}^{Q} \lambda_q g_{ai}^q\right) \\ &= -\lambda_0 \frac{dN}{dt} - \sum_{q=1}^{Q} \lambda_q \frac{d\hat{G}_q(t)}{dt} . \end{aligned} \tag{3.120}$$

By introducing the "socio-potential"

$$\hat{T}(t) = \hat{S} + \lambda_0 N + \sum_{q=1}^{Q} \lambda_q \hat{G}_q \tag{3.121}$$

there follows the equivalent formula

$$\frac{d\hat{T}(t)}{dt} = N\frac{d\lambda_0}{dt} + \sum_{q=1}^{Q} \hat{G}_q(t) \frac{d\lambda_q}{dt} . \tag{3.122}$$

The explicit time path follows by inserting (3.108) into the rhs of (3.120) or (3.122).

It should be noted that it is not possible to derive a "H-theorem", i.e. a monotic development of $\hat{S}(t)$ towards an equilibrium value, within the present framework of equations. The reason for this is that (3.100) and more so (3.108) are *approximate* equations for the mean values. As has been seen (for instance, in Chap. 2) the solutions of such equations may approach several stationary states depending on the initial conditions. Furthermore, the parameters $\boldsymbol{y}$ and $\boldsymbol{\kappa}$ may themselves be dynamic variables. In this case the equations (3.100) or (3.108) are coupled with the equations of motion of $\boldsymbol{y}$ and $\boldsymbol{\kappa}$. The solutions of the coupled system need not in general reach an equilibrium state.

The approximate closed equations for grossvariables derived in this section still have a rather complicated structure, since only relatively general assumptions have been used. In special models, however, it may be possible to find alternative and simpler derivations of grossvariable equations. Such an alternative procedure will be developed in Chap. 6 when setting up equations for certain grossvariables of political psychology in interacting societies.

4. Migration and/or Birth-Death Processes in Populations

In this chapter a model of migration and birth-death processes for interacting populations will be developed which may be considered as a typical application of the general concepts of Chap. 3.

The *"aspect"* considered in this model is *habitation* and the possible *attitudes* to this aspect refer to "living in certain habitats or quarters". The *transition probabilities* therefore refer to births and deaths within or to migration between the quarters.

There exists an extensive literature on birth-death and migration processes of biological species, of which [4.1–8] is only a small selection. This literature includes stochastic models as well as mean value models. Although there is, of course, some overlap with these models – the approach introduced here contains several important new features:

a) The migration of animals between habitats is usually described as a linear diffusion process with constant transition rates between neighbouring sites. This description seems to be appropriate for modelling "random migration" [4.1, 9] of populations of animals.

For *human* populations, however, such a description is insufficient: Human decisions to move to another place, for instance, to another part of a city, could take into account the existing distribution of different populations in the city, i.e. the existing socio-configuration. Furthermore, these decisions usually depend on valuations of the quality of neighbouring or distant quarters. Thus it can be expected that the *human* migration process is *non-linear* (with transition rates depending on the existing socio-configuration) as well as *non-diffusionary*, i.e. transitions to non-neighbouring quarters can take place. In Sect. 4.2 it will be seen that already the simplest *non-linear* migration process considering only two interacting populations between two parts of a city can lead to a variety of effects such as ghetto formation or cyclic migration depending on the values of the parameters describing the psychology of interactions between the populations.

b) Even the often treated pure birth-death process contains some interesting but seldom investigated problems which will be treated in Sect. 4.3. In particular the problem of the extinction of small populations of animals and the effect of including multi-step processes in the fully stochastic model will be investigated.

c) Another mathematically interesting and biologically realistic model arises if the well known Volterra-Lotka model is combined with non-linear migration of the predatory and prey species between habitats (Sect. 4.4). Table 4.1 summarizes the cases treated in the forthcoming sections of this chapter.

Table 4.1. Preview of migration and birth-death processes treated in Chap. 4

Number of populations	Number of quarters or habitats	Number of migrating populations	Birth/death processes included?	Sect.
P	L	P	yes	4.1
2	2	2	no	4.2
1	1	0	yes	4.3
2	2	1	yes	4.4

4.1 *The General Model

In this section the master equation and the mean value equations for the general migration and birth-death process of interacting populations, proceeding along the lines of Chap. 3, will be set up.

Assumed are P distinct populations $\mathcal{P}_\alpha$, $(\alpha = 1, 2, \ldots, P)$ and a one-dimensional attitude space $\mathfrak{A}$, because the only aspect under consideration is "habitation". Assuming L possible quarters or habitats $i = 1, 2, \ldots, L$, for instance in a city, these correspond to the attitudes i "to live in quarter i". The socio-configuration can be represented as a point in the C-dimensional socio-configuration space $\mathfrak{C}$ where $C = PL$:

$$\boldsymbol{n} \equiv \{n_{\alpha i}\} \in \mathfrak{C} \tag{4.1}$$

where $\alpha = 1, 2, \ldots, P$, $i = 1, 2, \ldots, L$ and $n_{\alpha i}$ is the number of members of $\mathcal{P}_\alpha$ living in quarter i.

The contributions to the transition probability $w[\boldsymbol{k}, \boldsymbol{n}]$ of change from a socio-configuration $\boldsymbol{n} \in \mathfrak{C}$ to another $(\boldsymbol{n} + \boldsymbol{k}) \in \mathfrak{C}$ are introduced by assuming, for simplicity, individual migration and birth-death processes; this means the number of group members simultaneously undergoing transition is 1. Thus, in the notation of Chap. 3:

a) The transition probability $w_{ji}^\alpha[\boldsymbol{k}, \boldsymbol{n}]$ due to migration of members of population $\mathcal{P}_\alpha$ from quarter i to quarter $j\,(i \neq j)$ is:

$$w_{ji}^\alpha[\boldsymbol{k}, \boldsymbol{n}] = \begin{cases} n_{\alpha i}\mu_{ji}^\alpha(\boldsymbol{n}, \boldsymbol{\kappa}) \quad \text{for} \\ \boldsymbol{k} = \{\ldots 0 \ldots 1_{\alpha j} \ldots 0 \ldots (-1)_{\alpha i} \ldots 0 \ldots\} \\ 0 \quad \text{for all other} \quad \boldsymbol{k} \end{cases} \tag{4.2}$$

* This section can be omitted on a first reading.

where $\mu_{ji}^{\alpha}(\boldsymbol{n}, \boldsymbol{\kappa})$ is the corresponding individual transition probability depending in general on $\boldsymbol{n}$ and some trend parameters $\boldsymbol{\kappa}$.

b) The transition probabilities $w_{i+}^{\alpha}[\boldsymbol{k}, \boldsymbol{n}]$ and $w_{i-}^{\alpha}[\boldsymbol{k}, \boldsymbol{n}]$ with respect to birth and death, respectively, of members of population $\mathcal{P}_\alpha$ in quarter i are:

$$w_{i+}^{\alpha}[\boldsymbol{k}, \boldsymbol{n}] = \begin{cases} n_{\alpha i}\beta_i^{\alpha} & \text{for} \\ \boldsymbol{k} = \{\dots 0 \dots 1_{\alpha i}, \dots 0 \dots\} & \\ 0 & \text{for all other} \quad \boldsymbol{k} \end{cases} \tag{4.3 a}$$

$$w_{i-}^{\alpha}[\boldsymbol{k}, \boldsymbol{n}] = \begin{cases} n_{\alpha i}\delta_i^{\alpha} + n_{\alpha i}^2\gamma_i^{\alpha} & \text{for} \\ \boldsymbol{k} = \{\dots 0 \dots (-1)_{\alpha i}, \dots 0 \dots\} & \\ 0 & \text{for all other} \quad \boldsymbol{k} \end{cases} \tag{4.3 b}$$

where β_i^{α}, δ_i^{α} and γ_i^{α} are the corresponding individual birth, death and saturation rates. The definitions (4.3 a, b) can, if necessary, be generalized to take account of birth or death rates which are dependent on the existing configuration in a much more complicated way.

c) The Volterra-Lotka transition probabilities for the interaction of prey (δ) and predatory (β) populations of animals in habitat i are:

$$w_i^{\delta\beta}[\boldsymbol{k}, \boldsymbol{n}] = \begin{cases} n_{\delta i} n_{\beta i} \nu_{\delta\beta} & \text{for} \\ \boldsymbol{k} = \{\dots 0 \dots (-1)_{\delta i} \dots 0 \dots 1_{\beta i} \dots\} & \\ 0 & \text{for all other} \quad \boldsymbol{k}\,. \end{cases} \tag{4.4}$$

By definition all the transition probabilities introduced in a–c) are positive semi-definite.

In the general case the total transition probability is the sum over all contributions (4.2–4):

$$\begin{aligned} w[\boldsymbol{k}, \boldsymbol{n}] = {} & \sum_{i,j=1}^{L} \sum_{\alpha=1}^{P} w_{ji}^{\alpha}[\boldsymbol{k}, \boldsymbol{n}] \\ & + \sum_{i=1}^{L} \sum_{\alpha=1}^{P} (w_{i+}^{\alpha}[\boldsymbol{k}, \boldsymbol{n}] + w_{i-}^{\alpha}[\boldsymbol{k}, \boldsymbol{n}]) \\ & + \sum_{i=1}^{L} \sum_{\delta\beta} w_i^{\delta\beta}[\boldsymbol{k}, \boldsymbol{n}]\,. \end{aligned} \tag{4.5}$$

In the last term on the rhs of (4.5) the sum over (δ, β) extends over such pairs of species for which δ is the prey of predator β. Inserting (4.5) into (3.13) the master equation of the model is obtained in the general form

$$\frac{dp(\boldsymbol{n};t)}{dt} = \left(\frac{\partial p}{\partial t}\right)_{\mathrm{M}}(\boldsymbol{n};t) + \left(\frac{\partial p}{\partial t}\right)_{\mathrm{BD}}(\boldsymbol{n};t) + \left(\frac{\partial p}{\partial t}\right)_{\mathrm{VL}}(\boldsymbol{n};t) \tag{4.6}$$

where $p(\boldsymbol{n};t)$ is the probability of a given socio-configuration $\boldsymbol{n}$ at time t and the three terms on the rhs of (4.6) refer to migration, birth-death and Volterra-Lotka processes respectively arising from the three contributions (4.2–4) to $w[\boldsymbol{k},\boldsymbol{n}]$ in (4.5).

For the explicit form of these terms it is convenient to introduce "transition operators" $E_{\alpha i}$ acting on a function of the socio-configuration as follows:

$$E_{\alpha i}^{\pm 1} f(n_{11} \dots n_{\alpha i} \dots n_{PL}) = f(n_{11} \dots (n_{\alpha i} \pm 1) \dots n_{PL}) \,. \tag{4.7}$$

Inserting (4.5) with (4.2–4) into (3.14) and using the definition (4.7) the explicit forms for the terms of the rhs of the master equation (4.6) are obtained as:

$$\left(\frac{\partial p}{\partial t}\right)_{\mathrm{M}}(\boldsymbol{n};t) = \sum_{i,j=1}^{L}\sum_{\alpha=1}^{P}(E_{\alpha i}E_{\alpha j}^{-1}-1)[n_{\alpha i}\mu_{ji}^{\alpha}(\boldsymbol{n},\boldsymbol{\kappa})\,p(\boldsymbol{n};t)] \tag{4.8}$$

$$\begin{aligned}\left(\frac{\partial p}{\partial t}\right)_{\mathrm{BD}}(\boldsymbol{n};t) = \sum_{i=1}^{L}\sum_{\alpha=1}^{P}\{&(E_{\alpha i}^{-1}-1)[n_{\alpha i}\beta_i^{\alpha}p(\boldsymbol{n};t)]\\ &+(E_{\alpha i}-1)[(n_{\alpha i}\delta_{\alpha}^{i}+n_{\alpha i}^{2}\gamma_i^{\alpha})\,p(\boldsymbol{n};t)]\}\end{aligned} \tag{4.9}$$

and

$$\left(\frac{\partial p}{\partial t}\right)_{\mathrm{VL}}(\boldsymbol{n};t) = \sum_{i=1}^{L}\sum_{\delta,\beta}(E_{\delta i}E_{\beta i}^{-1}-1)[n_{\delta i}n_{\beta i}v_{\delta\beta}\,p(\boldsymbol{n};t)] \,. \tag{4.10}$$

Directly from the master equation (4.6) it is possible to derive equations of motion for the mean values[1] as defined by

$$\langle n_{\gamma k}\rangle_t = \sum_{\boldsymbol{n}} n_{\gamma k}\,p(\boldsymbol{n};t) \,. \tag{4.11}$$

To this end (4.6) is multiplied by $n_{\gamma k}$ and the sum over all configurations $\boldsymbol{n}$ is taken. Using transition formulas like

$$n_{\gamma k}E_{\alpha i}E_{\alpha j}^{-1}f(\boldsymbol{n}) \equiv E_{\alpha i}E_{\alpha j}^{-1}(n_{\gamma k}-\delta_{\gamma\alpha}\delta_{ki}+\delta_{\gamma\alpha}\delta_{kj})f(\boldsymbol{n}) \tag{4.12}$$

[1] In Chap. 3 the Fokker-Planck equation (3.32) and not the master equation was used for the derivation of the mean value equations.

derived from (4.7) the exact result

$$\frac{d\langle n_{\gamma k}\rangle_t}{dt}$$

$$= \left[\sum_{i=1}^{L} \langle n_{\gamma i}\mu^{\gamma}_{ki}(\boldsymbol{n}, \boldsymbol{\kappa})\rangle_t - \sum_{j=1}^{L} \langle n_{\gamma k}\mu^{\gamma}_{jk}(\boldsymbol{n}, \boldsymbol{\kappa})\rangle_t\right]$$

$$+ [(\beta^{\gamma}_k - \delta^{\gamma}_k)\langle n_{\gamma k}\rangle_t - \gamma^{\gamma}_k \langle n^2_{\gamma k}\rangle_t]$$

$$+ \left[\sum_{\delta} \nu_{\delta\gamma}\langle n_{\delta k} n_{\gamma k}\rangle_t - \sum_{\beta} \nu_{\gamma\beta}\langle n_{\gamma k} n_{\beta k}\rangle_t\right] \tag{4.13}$$

is obtained.

The three terms on the rhs of (4.13) refer to migration, birth-death and prey-predatory processes respectively. In the last term the first sum extends over species δ which are the prey of the predator γ while the second sum extends over the predator species β with prey γ.

For distributions $p(\boldsymbol{n}; t)$ with only one sharp peak the exact equations (4.13) can be substituted by the approximate closed set of equations for the mean values $\langle n_{\gamma k}\rangle_t \equiv \tilde{n}_{\gamma k}$

$$\frac{d\tilde{n}_{\gamma k}}{dt}$$

$$= \left[\sum_{i=1}^{L} \tilde{n}_{\gamma i}\mu^{\gamma}_{ki}(\tilde{\boldsymbol{n}}, \boldsymbol{\kappa}) - \sum_{j=1}^{L} \tilde{n}_{\gamma k}\mu^{\gamma}_{jk}(\tilde{\boldsymbol{n}}, \boldsymbol{\kappa})\right]$$

$$+ [(\beta^{\gamma}_k - \delta^{\gamma}_k)\tilde{n}_{\gamma k} - \gamma^{\gamma}_k \tilde{n}^2_{\gamma k}]$$

$$+ \left[\sum_{\delta} \nu_{\delta\gamma}\tilde{n}_{\delta k}\tilde{n}_{\gamma k} - \sum_{\beta} \nu_{\gamma\beta}\tilde{n}_{\gamma k}\tilde{n}_{\beta k}\right]. \tag{4.14}$$

In the same manner equations for the variances can also be derived. This will be done in the following sections where special cases of the general model will be explicitly investigated, as indicated in Table 4.1.

4.2 Migration of Two Interacting Populations Between Two Parts of a City

In this section birth-death processes and Volterra-Lotka processes will be neglected and the simple but nevertheless non-trivial case of two distinct, non-linearly interacting human populations $\mathcal{P}_\mu$, $\mathcal{P}_\nu$ settling in and migrating between two parts of a city [4.10, 11] will be modelled. The two populations considered are assumed to be only a small fraction of a main "background population".

The number of members of population $\mathcal{P}_\mu$ and $\mathcal{P}_\nu$ in the city area i ($i = 1, 2$) are denoted by m_i and n_i, respectively. The total number of the members of $\mathcal{P}_\mu$ and $\mathcal{P}_\nu$ is to be conserved, thus

$$(m_1 + m_2) = 2\overline{m} \quad \text{and} \quad (n_1 + n_2) = 2\overline{n} . \tag{4.15}$$

Therefore one needs only two instead of four relevant integer variables m, n to characterize the socio-configuration, putting:

$$\begin{aligned} m_1 &= \overline{m} + m \quad &\text{and} \quad m_2 &= \overline{m} - m \quad &\text{with} \quad -\overline{m} \le m \le \overline{m} \\ n_1 &= \overline{n} + n \quad &\text{and} \quad n_2 &= \overline{n} - n \quad &\text{with} \quad -\overline{n} \le n \le \overline{n} . \end{aligned} \tag{4.16}$$

Using the individual transition probabilities $\mu_{ji}^{\mu}(m, n)$ and $\mu_{ji}^{\nu}(m, n)$ for members of $\mathcal{P}_\mu$ and $\mathcal{P}_\nu$, respectively, moving from area i to area j of the city, the probabilities $w[k, l; m, n]$ for a transition from the socio-configuration (m, n) to the configuration $(m + k, n + l)$ are obtained according to (4.2) as

$$\left.\begin{aligned} w[-1, 0; m, n] &= (\overline{m} + m)\mu_{21}^{\mu}(m, n) \\ w[+1, 0; m, n] &= (\overline{m} - m)\mu_{12}^{\mu}(m, n) \\ w[0, -1; m, n] &= (\overline{n} + n)\mu_{21}^{\nu}(m, n) \\ w[0, +1; m, n] &= (\overline{n} - n)\mu_{12}^{\nu}(m, n) . \end{aligned}\right\} \tag{4.17}$$

4.2.1 Master Equation, Mean Value, Variance and Fokker-Planck Equations

The master equation corresponding to (4.6) for the probability distribution $p(m, n; t)$ over the *two relevant variables* m and n, using (4.17), is

$$\begin{aligned} \frac{dp(m, n; t)}{dt} & \\ = (E_m - 1)\{w[-1, 0; m, n]p(m, n; t)\} &+ (E_m^{-1} - 1)\{w[1, 0; m, n]p(m, n; t)\} \\ + (E_n - 1)\{w[0, -1; m, n]p(m, n; t)\} &+ (E_n^{-1} - 1)\{w[0, 1; m, n]p(m, n; t)\} . \end{aligned} \tag{4.18}$$

Applying the same technique as in the derivation of (4.13) from (4.6) the equations of motion for the mean values of m, n, m^2, mn and n^2 can be directly derived, where, of course, mean values $\langle f(m, n)\rangle_t$ of $f(m, n)$ are defined and denoted by

$$\langle f(m, n)\rangle_t = \sum_{m=-\overline{m}}^{\overline{m}} \sum_{n=-\overline{n}}^{\overline{n}} f(m, n)p(m, n; t) . \tag{4.19}$$

The exact equations

$$\left.\begin{aligned}\frac{d\langle m\rangle_t}{dt} &= \langle k_m(m,n)\rangle_t \\ \frac{d\langle n\rangle_t}{dt} &= \langle k_n(m,n)\rangle_t\end{aligned}\right\} \tag{4.20}$$

and

$$\left.\begin{aligned}\frac{d\langle m^2\rangle_t}{dt} &= 2\langle m\,k_m(m,n)\rangle_t + \langle q_m(m,n)\rangle_t \\ \frac{d\langle mn\rangle_t}{dt} &= \langle m\,k_n(m,n)\rangle_t + \langle n\,k_m(m,n)\rangle_t \\ \frac{d\langle n^2\rangle_t}{dt} &= 2\langle n\,k_n(m,n)\rangle_t + \langle q_n(m,n)\rangle_t\end{aligned}\right\} \tag{4.21}$$

are obtained after introducing the "drift coefficients"

$$\left.\begin{aligned}k_m(m,n) &= w[1,0;m,n] - w[-1,0;m,n] \\ k_n(m,n) &= w[0,1;m,n] - w[0,-1;m,n]\end{aligned}\right\} \tag{4.22}$$

and the "fluctuation coefficients"

$$\left.\begin{aligned}q_m(m,n) &= w[1,0;m,n] + w[-1,0;m,n] \\ q_n(m,n) &= w[0,1;m,n] + w[0,-1;m,n]\,.\end{aligned}\right\} \tag{4.23}$$

Assuming a distribution $p(m,n;t)$ with only *one* sharp peak around the mean values $\langle m\rangle_t \equiv \tilde{m}(t)$ and $\langle n\rangle_t \equiv \tilde{n}(t)$ approximate equations for (4.20) are obtained as

$$\left.\begin{aligned}\frac{d\tilde{m}}{dt} &\approx k_m(\tilde{m},\tilde{n}) \\ \frac{d\tilde{n}}{dt} &\approx k_n(\tilde{m},\tilde{n})\,.\end{aligned}\right\} \tag{4.24}$$

Furthermore, by expanding $k_m(m,n)$, $k_n(m,n)$, $q_m(m,n)$ and $q_n(m,n)$ in Taylor series around $(\tilde{m},\tilde{n})$ and considering only the lowest terms of the expansions for the rhs of (4.21), the approximate equations

$$\left.\begin{aligned}\frac{d\sigma_{mm}}{dt} &= \tilde{q}_m + 2\sigma_{mm}\partial_m\tilde{k}_m + 2\sigma_{mn}\partial_n\tilde{k}_m\\ \frac{d\sigma_{mn}}{dt} &= \sigma_{mm}\partial_m\tilde{k}_n + \sigma_{mn}(\partial_n\tilde{k}_n + \partial_m\tilde{k}_m) + \sigma_{nn}\partial_n\tilde{k}_m\\ \frac{d\sigma_{nn}}{dt} &= \tilde{q}_n + 2\sigma_{nn}\partial_n\tilde{k}_n + 2\sigma_{nm}\partial_m\tilde{k}_n\end{aligned}\right\} \tag{4.25}$$

are obtained where

$$\partial_m \equiv \partial/\partial_m,\ \partial_n \equiv \partial/\partial_n,\ \tilde{k} = k(\tilde{m}, \tilde{n}) \qquad \text{and} \quad \tilde{q} = q(\tilde{m}, \tilde{n})$$

and variances are denoted by

$$\left.\begin{aligned}\sigma_{mm} &= \langle m^2\rangle_t - \langle m\rangle_t^2\\ \sigma_{mn} &= \langle mn\rangle_t - \langle m\rangle_t\langle n\rangle_t\\ \sigma_{nn} &= \langle n^2\rangle_t - \langle n\rangle_t^2 .\end{aligned}\right\} \tag{4.26}$$

For convenience, the normalized variables x, y instead of m, n are now used:

$$x = \frac{m}{\overline{\overline{m}}} \quad \text{and} \quad y = \frac{n}{\overline{\overline{n}}} \quad \text{where} \quad -1 \le x, y \le +1 \tag{4.27}$$

and the drift and fluctuation coefficients will be written in the form

$$\left.\begin{aligned}k_m(m, n) &= \overline{m}K_m(x, y) \quad \text{and} \quad k_n(m, n) = \overline{n}K_n(x, y)\\ q_m(m, n) &= \overline{m}Q_m(x, y) \quad \text{and} \quad q_n(m, n) = \overline{n}Q_n(x, y) .\end{aligned}\right\} \tag{4.28}$$

Writing x, y for $\tilde{x}$, $\tilde{y}$ the mean value equations (4.24) then take the form

$$\frac{dx}{dt} = K_m(x, y) \quad \text{and} \quad \frac{dy}{dt} = K_n(x, y) . \tag{4.29}$$

After introducing

$$\left.\begin{aligned}\sigma_{xx} &= \frac{1}{\overline{\overline{m}}^2}\sigma_{mm} = \langle(x - \langle x\rangle)^2\rangle\\ \sigma_{xy} &= \frac{1}{\overline{\overline{m}}\,\overline{\overline{n}}}\sigma_{mn} = \langle(x - \langle x\rangle)(y - \langle y\rangle)\rangle\\ \sigma_{yy} &= \frac{1}{\overline{\overline{n}}^2}\sigma_{nn} = \langle(y - \langle y\rangle)^2\rangle\end{aligned}\right\} \tag{4.30}$$

instead of σ_{mm}, σ_{mn} and σ_{nn}, the variance equations (4.25) appear as

$$\left.\begin{aligned}\frac{d\sigma_{xx}}{dt} &= \frac{1}{\bar{m}} Q_m + 2\sigma_{xx}\partial_x K_m + 2\sigma_{xy}\partial_y K_m \\ \frac{d\sigma_{xy}}{dt} &= \sigma_{xx}\partial_x K_n + \sigma_{xy}(\partial_x K_m + \partial_y K_n) + \sigma_{yy}\partial_y K_m \\ \frac{d\sigma_{yy}}{dt} &= \frac{1}{\bar{n}} Q_n + 2\sigma_{yy}\partial_y K_n + 2\sigma_{yx}\partial_x K_n \,.\end{aligned}\right\} \tag{4.31}$$

The procedure in Sect. 3.2.2 leading from the master equation (3.14) to the Fokker-Planck equation (3.32) yields for the present model the Fokker-Planck equation

$$\begin{aligned}&\frac{\partial P(x, y; t)}{\partial t} \\ &\quad = -\{\partial_x[K_m(x, y)P(x, y; t)] + \partial_y[K_n(x, y)P(x, y; t)]\} \\ &\quad + \frac{1}{2\bar{m}}\partial_x^2[Q_m(x, y)P(x, y; t)] + \frac{1}{2\bar{n}}\partial_y^2[Q_n(x, y)P(x, y; t)]\end{aligned} \tag{4.32}$$

for the probability distribution $P(x, y; t)$, which is normalized by means of

$$\int_{-1}^{+1}\int_{-1}^{+1} dx\, dy\, P(x, y; t) = 1 \,. \tag{4.33}$$

From the definitions (4.17, 22, 23, 28) it follows that in general the drift and fluctuation coefficients in (4.32) depend on the variables x and y.

Furthermore it is easily checked that the mean value and variance equations derived from (4.32) by proceeding as in Sect. 3.2.4 agree exactly with the results (4.29, 31) derived directly from the master equation (4.18).

In order to make the model as developed so far fully explicit as well as applicable, the individual transition probabilities $\mu_{ji}^{\mu}(m, n)$ and $\mu_{ji}^{\nu}(m, n)$ of the members of the two populations $\mathcal{P}_\mu$, $\mathcal{P}_\nu$ have to be adequately modelled. The model adopted, which is sufficiently flexible in that several possible effects on the migration process of the psychology of interacting populations are included, is given by

$$\left.\begin{aligned}\mu_{12}^{\mu}(m, n) &= \gamma\exp[u(m, n)] &&\equiv \gamma\exp\{u[x, y]\} \\ \mu_{21}^{\mu}(m, n) &= \gamma\exp[-u(m, n)] &&\equiv \gamma\exp\{-u[x, y]\} \\ \mu_{12}^{\nu}(m, n) &= \gamma\exp[v(m, n)] &&\equiv \gamma\exp\{v[x, y]\} \\ \mu_{21}^{\nu}(m, n) &= \gamma\exp[-v(m, n)] &&\equiv \gamma\exp\{-v[x, y]\}\end{aligned}\right\} \tag{4.34}$$

where

$$\left.\begin{aligned} u(m,n) &= (\pi_\mu + \hat{\kappa}_\mu m + \hat{\sigma}_\mu n) \\ &= u[x,y] = (\pi_\mu + \kappa_\mu x + \sigma_\mu y) \\ \text{and} \\ v(m,n) &= (\pi_\nu + \hat{\kappa}_\nu n + \hat{\sigma}_\nu m) \\ &= v[x,y] = (\pi_\nu + \kappa_\nu y + \sigma_\nu x) \end{aligned}\right\} \quad (4.35)$$

with

$$\left.\begin{aligned} \kappa_\mu &= \overline{m}\,\hat{\kappa}_\mu \quad \text{and} \quad \sigma_\mu = \overline{n}\,\hat{\sigma}_\mu \\ \kappa_\nu &= \overline{n}\,\hat{\kappa}_\nu \quad \text{and} \quad \sigma_\nu = \overline{m}\,\hat{\sigma}_\nu\,. \end{aligned}\right\} \quad (4.36)$$

An interpretation of the parameters γ, π_μ, π_ν, κ_μ, κ_ν, σ_μ and σ_ν can easily be deduced by examining (4.34) with (4.35) in connection with (4.17). In general, the individual transition probabilities characterize the degree of readiness of the members of the populations $\mathcal{P}_\mu$ and $\mathcal{P}_\nu$ to move to the other half or area of the city where the existing number of members of $\mathcal{P}_\mu$ and $\mathcal{P}_\nu$ in both areas of the city and the qualitative "value" of these areas are taken into account. As the interpretation of γ, π_ν, κ_ν, σ_ν is synonymous with that of γ, π_μ, κ_μ, σ_μ, only the set γ, π_μ, κ_μ, σ_μ need be discussed.

The parameter γ determines the general time scale on which migration processes take place and is therefore denoted as *the flexibility parameter*.

The parameter π_μ describes the level of preference given to one of the two city areas by the members of $\mathcal{P}_\mu$: For $\pi_\mu > 0$ ($\pi_\mu < 0$) area 1 (area 2) is preferred by members of $\mathcal{P}_\mu$. π_μ is therefore denoted as *the preference parameter of* $\mathcal{P}_\mu$.

The parameter κ_μ (> 0 or < 0) describes the strength of the (positive or negative) trend for members of $\mathcal{P}_\mu$ to cluster or group in the same area of the city, i.e. is the level of the trend towards "living together" for this population group. κ_μ is therefore denoted as *the internal sympathy parameter* for $\mathcal{P}_\mu$.

The parameter σ_μ (> 0 or < 0) describes the strength of the (positive or negative) trend for members of $\mathcal{P}_\mu$ to live together with members of the population $\mathcal{P}_\nu$ in the same area of the city. σ_μ is therefore denoted as *the external sympathy parameter* for $\mathcal{P}_\mu$.

In (4.34) the several trends described by π_μ, κ_μ and σ_μ are automatically superimposed on each other. All trend parameters are assumed to remain constant in the time interval in which the migration process is to be investigated. In principle, however, it is also possible to take into account a time dependence for all trend parameters (Sect. 3.3).

Despite their suggestive denotations it is *neither intended nor necessary* to give an explicit psychological explanation of the motivation giving rise to these parameters. It is sufficient for the validity of the model that any bundle of individual motivations leads to migration trends which can be described by the transition probabilities (4.34) or by their possible further generalized forms.

4.2.2 Solutions of the Equations in Relevant Cases

If (4.17, 34, 35) are incorporated into the *master equation* (4.18) and by using (4.22, 23, 28) in the *mean value* and *variance equations* (4.29, 31) as well as in the *Fokker-Planck equation* (4.32), the explicit form for all kinds of equations of motion is obtained. Relevant cases for which solutions can be obtained either analytically or by numerical methods will now be considered.

Firstly the *stationary solution* $p_{st}(m, n)$ to the master equation has to be sought. This solution takes into account the most probable configurations (m, n) and the fluctuations around them in a stationary situation established by the migration process after the relaxation of a possible initial imbalance.

Haken [4.12] has derived a form for the solution $p_{st}(\boldsymbol{i})$ to the general stationary master equation, i.e. to the equation

$$0 = \frac{dp_{st}(\boldsymbol{i})}{dt} = \sum_{j} [w(\boldsymbol{j}, \boldsymbol{i})\, p_{st}(\boldsymbol{i}) - w(\boldsymbol{i}, \boldsymbol{j})\, p_{st}(\boldsymbol{j})] \tag{4.37}$$

where $\boldsymbol{i}$ and $\boldsymbol{j}$ can be multidimensional indices, in the case of *"detailed balance"*, i.e. if the transition probabilities $w(\boldsymbol{j}, \boldsymbol{i})$ from $\boldsymbol{i}$ to $\boldsymbol{j}$ fulfil the condition

$$w(\boldsymbol{j}, \boldsymbol{i})\, p_{st}(\boldsymbol{i}) = w(\boldsymbol{i}, \boldsymbol{j})\, p_{st}(\boldsymbol{j}) \tag{4.38}$$

for every index pair $(\boldsymbol{j}, \boldsymbol{i})$. The form of the solution, given below as (4.39), is intuitively apparent: Starting from an arbitrarily chosen site denoted by $\boldsymbol{i}_0 = \boldsymbol{0}$ there has to exist at least one chain $\mathscr{C}$, in general many chains, of sites

$$\boldsymbol{i}_1 \equiv \boldsymbol{1},\ \boldsymbol{i}_2 \equiv \boldsymbol{2}, \ldots, \boldsymbol{i}_{n-1} \equiv (\boldsymbol{n} - \boldsymbol{1}),\ \boldsymbol{i}_n \equiv \boldsymbol{j}$$

with non-vanishing transition probabilities $w(\boldsymbol{1}, \boldsymbol{0})$, $w(\boldsymbol{0}, \boldsymbol{1})$, $w(\boldsymbol{2}, \boldsymbol{1})$, $w(\boldsymbol{1}, \boldsymbol{2}), \ldots, w(\boldsymbol{j}, \boldsymbol{n} - \boldsymbol{1})$, $w(\boldsymbol{n} - \boldsymbol{1}, \boldsymbol{j})$ to every other site $\boldsymbol{j} \neq \boldsymbol{0}$. If this is not so, $\boldsymbol{j}$ is an isolated site never reached by a probability flux if started from an initial distribution $p(\boldsymbol{i}; \boldsymbol{0}) = \delta_{\boldsymbol{i}\boldsymbol{0}}$. Considering one of these chains, the stationary solution $p_{st}(\boldsymbol{j})$ is obtained by repeated application of (4.38) in the form

$$p_{st}(\boldsymbol{j}) = p_{st}(\boldsymbol{0}) \prod_{\boldsymbol{v}=\boldsymbol{0}}^{\boldsymbol{n}-\boldsymbol{1}} \frac{w(\boldsymbol{v} + \boldsymbol{1}, \boldsymbol{v})}{w(\boldsymbol{v}, \boldsymbol{v} + \boldsymbol{1})}\ . \tag{4.39}$$

By using (4.38) this result can be proved to be unique and therefore *independent* of the choice of the chain $\mathscr{C}$. The normalization of the solution (4.39) finally determines the magnitude of $p_{st}(\boldsymbol{0})$.

Simultaneously conditions for $w(\boldsymbol{j}, \boldsymbol{i})$ equivalent to (4.38) but not containing the stationary solution can be derived from (4.39): For any closed loop

$\mathscr{L}(\boldsymbol{i}=\mathbf{0}, \mathbf{1}, \mathbf{2}, \ldots, (\boldsymbol{l}-\mathbf{1}), \quad \boldsymbol{l} \equiv \mathbf{0})$ of sites with $w(\boldsymbol{i}+\mathbf{1}, \boldsymbol{i}) \neq \mathbf{0}$ and $w(\boldsymbol{i}, \boldsymbol{i}+\mathbf{1}) \neq \mathbf{0}$ and by applying (4.39) for $\boldsymbol{j}=\mathbf{0}$ and $\mathscr{C}$ substituted by $\mathscr{L}$

$$(\mathscr{L}) \prod_{\boldsymbol{\nu}=\mathbf{0}}^{\boldsymbol{l}-\mathbf{1}} \frac{w(\boldsymbol{\nu}+\mathbf{1}, \boldsymbol{\nu})}{w(\boldsymbol{\nu}, \boldsymbol{\nu}+\mathbf{1})}=1 \tag{4.40}$$

is obtained. Eq. (4.40) is equivalent to (4.38) and formulates the condition for detailed balance without using the stationary solution $p_{\mathrm{st}}(\boldsymbol{i})$.

The model cases for which the condition of detailed balance is fulfilled will now be checked. As there exist transition probabilities (4.17) only between *neighbouring sites* in the (m, n) – space, the necessary and sufficient condition for detailed balance is that (4.40) holds good for all elementary closed loops

$$\mathscr{L}((m, n); (m+1, n); (m+1, n+1); (m, n+1); (m, n)) .$$

This implies that

$$\frac{w[1,0; m n]\, w[0,1; m+1, n]\, w[-1,0; m+1, n+1]\, w[0,-1; m, n+1]}{w[-1,0; m+1, n]\, w[0,-1; m+1, n+1]\, w[1,0; m, n+1]\, w[0,1; m, n]} = 1 . \tag{4.41}$$

Inserting (4.34, 35) in (4.17, 41) yields the condition

$$\left.\begin{array}{ll} \exp[2(\hat{\sigma}_\mu-\hat{\sigma}_\nu)]=1 & \text{or} \\ \hat{\sigma}_\mu=\hat{\sigma}_\nu \equiv \hat{\sigma} & \text{and} \quad \bar{m}\sigma_\mu=\bar{n}\sigma_\nu \end{array}\right\} \tag{4.42}$$

with the implication that detailed balance only holds good for equal external sympathy parameters. In this case the exact stationary solution to (4.18) can be set up by using (4.39) and by choosing, for example, the chain

$$\mathscr{C}((0,0), (1,0), \ldots, (m, 0); (m, 1), \ldots, (m, n)) .$$

Then

$$p_{\mathrm{st}}(m, n)=p_{\mathrm{st}}(0,0) \prod_{\mu=0}^{m-1} \frac{w[1,0; \mu, 0]}{w[-1,0; \mu+1, 0]} \prod_{\nu=0}^{n-1} \frac{w[0,1; m, \nu]}{w[0,-1; m, \nu+1]} . \tag{4.43}$$

Inserting (4.34,35) in (4.17) and taking the products, the exact result

$$p_{\mathrm{st}}(m, n) = C\binom{2\bar{m}}{\bar{m}+m}\binom{2\bar{n}}{\bar{n}+n} \exp[\varphi(m, n)] \tag{4.44}$$

is obtained where

$$\varphi(m, n) = (2\pi_\mu m + 2\pi_\nu n + \hat{\kappa}_\mu m^2 + 2\hat{\sigma} mn + \hat{\kappa}_\nu n^2) \quad (4.45)$$

and C is the normalization factor. An evaluation of the binomial coefficients using *Stirling's* formula finally leads to

$$p_{st}(m, n) \approx C' \exp[\Phi(m, n)] \quad (4.46)$$

with

$$\begin{aligned}\Phi(m, n) &= \varphi(m, n)\\ &-[(\overline{m} + m)\ln(\overline{m} + m) + (\overline{m} - m)\ln(\overline{m} - m)]\\ &-[(\overline{n} + n)\ln(\overline{n} + n) + (\overline{n} - n)\ln(\overline{n} - n)]\ .\end{aligned} \quad (4.47)$$

The extrema of the distribution (4.46) are determined by the equations

$$\left.\begin{aligned}\frac{\partial\Phi(m, n)}{\partial m} &= 0, \quad \text{or} \quad \frac{m}{\overline{m}} \equiv x = \tanh(u[x, y])\\ \frac{\partial\Phi(m, n)}{\partial n} &= 0, \quad \text{or} \quad \frac{n}{\overline{n}} \equiv y = \tanh(v[x, y]\end{aligned}\right\} \quad (4.48)$$

The case where (4.42) does *not* hold, i.e. where detailed balance is not attained, will now be considered. In this case only rather complicated representations of the stationary solution of (4.18) are generally available [4.13, 14].

However, the form (4.44) of the solution *with* detailed balance suggests the following formulation for the case *without* detailed balance

$$p_{st}(m, n) = C\binom{2\overline{m}}{\overline{m} + m}\binom{2\overline{n}}{\overline{n} + n}\exp[\chi(m, n)] \quad (4.49)$$

where $\chi(m, n)$ is approximated by a polynomial of second order in m, n. Inserting (4.49) into the stationary master equation (4.18), expanding the exponential expressions in the appropriate way and comparing coefficients leads to the determination of $\chi(m, n)$ [4.10] as

$$\chi(m, n) = 2\pi_\mu m + 2\pi_\nu n + \hat{\kappa}_\mu m^2 + (\hat{\delta}_\mu + \hat{\sigma}_\nu) mn + \hat{\kappa}_\nu n^2\ . \quad (4.50)$$

The form of $p_{st}(m, n)$ in (4.49) together with (4.50) clearly comprises (4.44) but extends it, as a first approximation, to the case *without* detailed balance.

The non-stationary migration process will now be examined. Starting from a probability distribution $P(x, y; t = 0)$ with one sharp peak only, it seems reasonable to assume that the distribution *remains uni-modal* over a considerable period of time. As long as this assumption is correct the approximate but closed mean value equations (4.29) and variance equations (4.31) already contain all the relevant information. The solutions shown in Fig. 4.8 c support the validity of this assumption.

The explicit form of the relevant equations comes from (4.29) and (4.31) by inserting the explicit drift and fluctuation coefficients (4.28), using (4.17, 34) in the definitions (4.22, 23)

$$\left.\begin{aligned} K_m(x, y) &= 2\gamma\{\sinh(u[x, y]) - x\cosh(u[x, y])\} \\ K_n(x, y) &= 2\gamma\{\sinh(v[x, y]) - y\cosh(v[x, y])\} \end{aligned}\right\} \tag{4.51}$$

$$\left.\begin{aligned} Q_m(x, y) &= 2\gamma\{\cosh(u[x, y]) - x\sinh(u[x, y])\} \\ Q_n(x, y) &= 2\gamma\{\cosh(v[x, y]) - y\sinh(v[x, y])\}\ . \end{aligned}\right\} \tag{4.52}$$

Thus, using (4.29, 51) and introducing from now on a scaled dimensionless time $\tau = 2\gamma t$, the following explicit mean value equations (writing x, y instead of $\bar{x}$, $\bar{y}$ for the mean values) are obtained as

$$\left.\begin{aligned} &\frac{dx}{d\tau} = [\tanh(u) - x]\cosh(u) \\ &\frac{dy}{d\tau} = [\tanh(v) - y]\cosh(v) \qquad \text{where} \\ &u = (\pi_\mu + \kappa_\mu x + \sigma_\mu y) \qquad \text{and} \quad v = (\pi_\nu + \kappa_\nu y + \sigma_\nu x)\ . \end{aligned}\right\} \tag{4.53}$$

The explicit but lengthy variance equations obtained analogously from (4.31, 51, 52) need not be derived. It should be pointed out, however, that for vanishing preference parameters, i.e. for $\pi_\mu = \pi_\nu = 0$, the equations (4.53) are invariant under the transformations

a) $x \to -x\ ; \quad y \to -y$

b) $x \to x\ ; \quad y \to -y \quad \sigma_\nu \to -\sigma_\nu\ ; \quad \sigma_\mu \to -\sigma_\mu\ .$

Invariance a) reflects the equivalence of both parts of the city for $\pi_\mu = \pi_\nu = 0$. Invariance b) implies, for instance, that to a solution for which $x(\tau) > 0$ and $y(\tau) > 0$ concerning populations with *positive* external sympathy parameters, where $\sigma_\nu > 0$ and $\sigma_\mu > 0$ describe a *clustering* of both populations $\mathcal{P}_\nu$, $\mathcal{P}_\mu$ in the same area of the city, there exists a corresponding solution $x'(\tau) = x(\tau)$ and $y'(\tau) = -y(\tau)$ for populations with *negative* external sympathy parameters $\sigma'_\nu = -\sigma_\mu$ and $\sigma'_\mu = -\sigma_\mu$ describing a *separation* of the two populations.

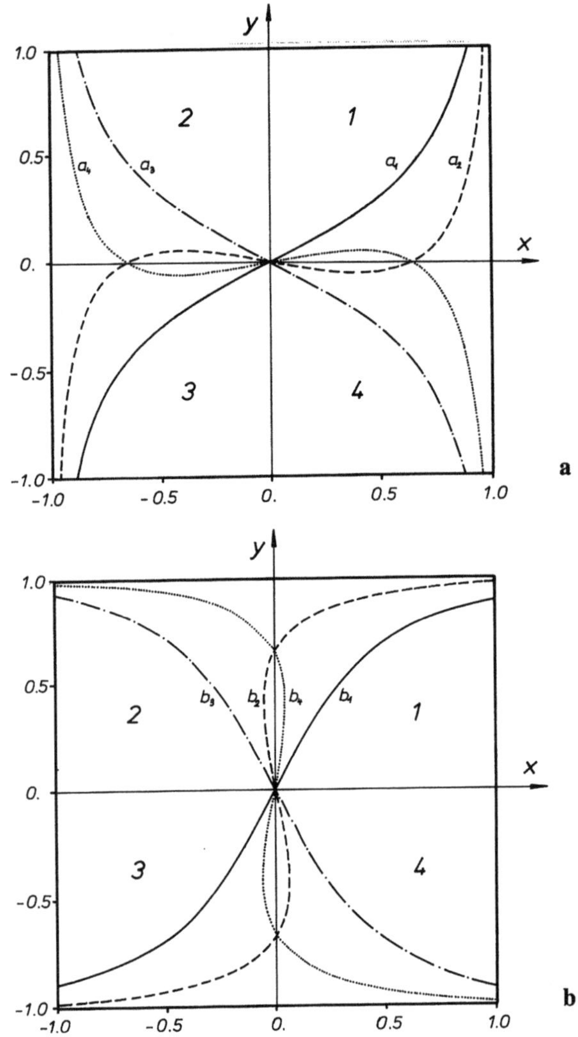

Figure 4.1. **(a)** Typical graphs of the function $F_\nu(x, y) = 0$ for the cases $a_1 (0 < \kappa_\mu < 1, \sigma_\mu > 0)$, $a_2 (\kappa_\mu > 1, \sigma_\mu > 0)$, $a_3 (0 < \kappa_\mu < 1, \sigma_\mu < 0)$ and $a_4 (\kappa_\mu > 1, \sigma_\mu < 0)$
(b) Typical graphs of the function $F_\nu(x, y) = 0$ for the cases $b_1 (0 < \kappa_\nu < 1, \sigma_\nu > 0)$, $b_2 (\kappa_\nu > 1, \sigma_\nu > 0)$, $b_3 (0 < \kappa_\nu < 1, \sigma_\nu < 0)$ and $b_4 (\kappa_\nu > 1, \sigma_\nu < 0)$

The global structure of the change in mean values with time of the migration system depends decisively on the *stationary* or *singular points* $P_j(x_j, y_j)$ of the differential equations (4.53). These are defined by

$$\left.\begin{aligned} \left(\frac{dx}{d\tau}\right)_{P_j} &= \cosh(u_j)[\tanh(u_j) - x_j] = 0 \\ \left(\frac{dy}{d\tau}\right)_{P_j} &= \cosh(v_j)[\tanh(v_j) - y_j] = 0 \end{aligned}\right\} \qquad (4.54)$$

or, equivalently, by

$$\left.\begin{aligned} F_\mu(x_j, y_j) &\equiv \{\tanh(u[x_j, y_j]) - x_j\} = 0 \\ F_\nu(x_j, y_j) &\equiv \{\tanh(v[x_j, y_j]) - y_j\} = 0 . \end{aligned}\right\} \tag{4.55}$$

The functions $F_\mu(x, y) = 0$ and $F_\nu(x, y) = 0$ are plotted in the $x - y$ plane in Figs. 4.1 a, b for four typical choices of the trend parameters contained in the functions $u[x, y]$ and $v[x, y]$.

The singular points $P_j(x_j, y_j)$ for given trend parameters $(\pi_\mu, \kappa_\mu, \sigma_\mu; \pi_\nu, \kappa_\nu, \sigma_\nu)$ are intersection points of a graph of $F_\mu(x, y) = 0$ belonging to $(\pi_\mu, \kappa_\mu, \sigma_\mu)$ with a graph of $F_\nu(x, y) = 0$ belonging to $(\pi_\nu, \kappa_\nu, \sigma_\nu)$. In the symmetrical case, (i.e. for $\pi_\mu = \pi_\nu = 0$) there can exist 1, 3, 5 or 9 singular points $P_j(x_j, y_j)$ depending on the choice of the trend parameters.

A so-called *linear stability analysis* now answers the question whether a point $P(x(\tau), y(\tau))$ in the vicinity of a singular point $P_j(x_j, y_j)$ moving according to (4.53) will *approach* (or *move away from*) P_j. In the first (second) case P_j is denoted as a stable (unstable) focus.

For this analysis small deviations from the singular point $P_j(x_j, y_j)$ are introduced

$$\xi(\tau) = x(\tau) - x_j \quad \text{and} \quad \eta(\tau) = y(\tau) - y_j \tag{4.56}$$

and (4.53) is linearized in $\xi(\tau)$, $\eta(\tau)$ as

$$\left.\begin{aligned} \frac{d\xi}{d\tau} &= -\gamma_{\mu j}\xi + \sigma_{\mu j}\eta \\ \frac{d\eta}{d\tau} &= \sigma_{\nu j}\xi - \gamma_{\nu j}\eta \end{aligned}\right\} \tag{4.57}$$

where

$$\left.\begin{aligned} \kappa_{\mu j} &= \frac{\kappa_\mu}{\cosh(u_j)} && \text{and} \quad \kappa_{\nu j} = \frac{\kappa_\nu}{\cosh(v_j)} \\ \gamma_{\mu j} &= [\cosh(u_j) - \kappa_{\mu j}] && \text{and} \quad \gamma_{\nu j} = [\cosh(v_j) - \kappa_{\nu j}] \\ \sigma_{\mu j} &= \frac{\sigma_\mu}{\cosh(u_j)} && \text{and} \quad \sigma_{\nu j} = \frac{\sigma_\nu}{\cosh(v_j)} . \end{aligned}\right\} \tag{4.58}$$

The solutions to (4.57) are linear combinations of the eigensolutions:

$$\xi_\pm(\tau) = \xi_{0\pm}\exp(\lambda_{j\pm}\tau) \quad \text{and} \quad \eta_\pm(\tau) = \eta_{0\pm}\exp(\lambda_{j\pm}\tau) \tag{4.59}$$

with eigenvalues:

$$\left.\begin{aligned}\lambda_{j\pm} &= -\frac{1}{2}\left[(\gamma_{\mu j}+\gamma_{\nu j}) \pm \sqrt{(\gamma_{\mu j}+\gamma_{\nu j})^2+4\,\varepsilon_j}\right]\\ &\qquad \text{where}\quad \varepsilon_j \equiv \sigma_{\mu j}\sigma_{\nu j}-\gamma_{\mu j}\gamma_{\nu j}\,.\end{aligned}\right\} \tag{4.60}$$

The eigenvalues λ_{j+} and λ_{j-} are either real or conjugate complex numbers. Clearly P_j is a stable focus if $\mathrm{Re}\{\lambda_{j+}\} < 0$ and $\mathrm{Re}\{\lambda_{j-}\} < 0$ while P_j is an unstable focus if $\mathrm{Re}\{\lambda_{j+}\} > 0$ and/or $\mathrm{Re}\{\lambda_{j-}\} > 0$. If $\lambda_{j\pm}$ are real (complex), $P(\tau)$ approaches or moves away from P_j linearly (spirally). Following (4.60) Table 4.2 shows a simple delineation into five main cases.

Table 4.2. The main cases of the stability analysis

Conditions for parameters	Eigenvalues	P_j
Case a)		
$\gamma_{\mu j}+\gamma_{\nu j} > 0$	$\lambda_{j\pm}$ are conjugate complex	stable
$-\infty < \varepsilon_j < -\frac{1}{2}(\gamma_{\mu j}+\gamma_{\nu j})$	$\mathrm{Re}\{\lambda_{j\pm}\} < 0$	
Case b)		
$\gamma_{\mu j}+\gamma_{\nu j} > 0$	$\mathrm{Im}\{\lambda_{j\pm}\} = 0$	stable
$-\frac{1}{2}(\gamma_{\mu j}+\gamma_{\nu j}) < \varepsilon_j < 0$	$\lambda_{j\pm} < 0$	
Case c)		
$\gamma_{\mu j}+\gamma_{\nu j} < 0$	$\lambda_{j\pm}$ are conjugate complex	unstable
$-\infty < \varepsilon_j < -\frac{1}{2}\lvert\gamma_{\mu j}+\gamma_{\nu j}\rvert$	$\mathrm{Re}\{\lambda_{j\pm}\} > 0$	
Case d)		
$\gamma_{\mu j}+\gamma_{\nu j} < 0$	$\mathrm{Im}\{\lambda_{j\pm}\} = 0$	unstable
$-\frac{1}{2}\lvert\gamma_{\mu j}+\gamma_{\nu j}\rvert < \varepsilon_j < 0$	$\lambda_{j\pm} > 0$	
Case e)		
$\gamma_{\mu j}+\gamma_{\nu j} \gtreqless 0$	$\mathrm{Im}\{\lambda_{j\pm}\} = 0$	unstable
$\varepsilon_j > 0$	$\lambda_{j+}\lambda_{j-} < 0$	

A case of particular interest arises under the conditions

$$\kappa_\mu+\kappa_\nu > 2,\quad \sigma_\mu\sigma_\nu < 0 \quad\text{and}\quad \pi_\mu = \pi_\nu = 0 \tag{4.61}$$

for which the origin $P(0, 0)$ is the only singular point being then an unstable focus of the kind Case c) (Table 4.2). Furthermore, the flux lines of (4.53) in the domain $\mathfrak{D}(-1 \leq x \leq +1;\ -1 \leq y \leq +1)$ are always directed inward at the boundary because

$$\left.\begin{array}{llll} \dfrac{dx}{d\tau} < 0 & \text{for } x = 1 & \text{and } \dfrac{dx}{d\tau} > 0 & \text{for } x = -1 \\[2ex] \dfrac{dy}{d\tau} < 0 & \text{for } y = 1 & \text{and } \dfrac{dy}{d\tau} > 0 & \text{for } y = -1\,. \end{array}\right\} \quad (4.62)$$

This means that in the domain $\mathfrak{D}^*$ consisting of $\mathfrak{D}$ but without the origin $P(0, 0)$, there must exist a *limit cycle* as solution of (4.53) according to the *Poincaré-Bendixon* theorem [4.2, 12]. The solution computed and shown in Fig. 4.8 a (to be explained later) verifies this statement.

Some sets of reasonable trend parameters will now be chosen and the resulting paths of the mean values $x(\tau)$, $y(\tau)$ graphed and discussed (Fig. 4.2 a–8 a). In one case of particular interest the variances $\sigma_{xx}(\tau)$, $\sigma_{xy}(\tau)$ and $\sigma_{yy}(\tau)$ will also be considered (Fig. 4.8 c). All the figures result from computed solutions to (4.53 and 31) for the respective choices for the trend parameters. In addition, the solutions (4.44, 49) to the stationary master equation for the same trend parameters are shown in Figs. 4.2 b–8 b. For illustrative purposes $\overline{m}$ and $\overline{n}$ are both given the low value 20 in all calculations which leads to a relatively large width for the resulting distributions.

In interpreting all the following figures it has to be kept in mind that according to the definition of the variables x, y the subdomains of $\mathfrak{D}$ belong to the following situations:

$\mathfrak{D}_{++}(0 < x < 1$	and $0 < y < 1)$:	Both populations concentrate in city area 1
$\mathfrak{D}_{-+}(-1 < x < 0$	and $0 < y < 1)$:	$\mathcal{P}_\mu$ concentrates in area 2 and $\mathcal{P}_\nu$ concentrates in area 1
$\mathfrak{D}_{+-}(0 < x < 1$	and $-1 < y < 0)$:	$\mathcal{P}_\mu$ concentrates in area 1 and $\mathcal{P}_\nu$ concentrates in area 2
$\mathfrak{D}_{--}(-1 < x < 0$	and $-1 < y < 0)$:	Both populations concentrate in city area 2

For the results shown in Figs. 4.2 a, b weak internal and external sympathy parameters are assumed. Figure 4.2 a shows that all mean values trend into the stable focus $P(0, 0)$. This focus represents a homogeneous distribution of both populations over both city areas. Correspondingly the stationary probability distribution shown in Fig. 4.2 b concentrates around this stable focus.

This result is intuitively obvious: The question of how many members of both populations live in area 1 and 2 of the city does not play a prominent role in the migration decisions for members of populations with *weak* internal and external sympathy parameters. Furthermore, no preference for one of the areas

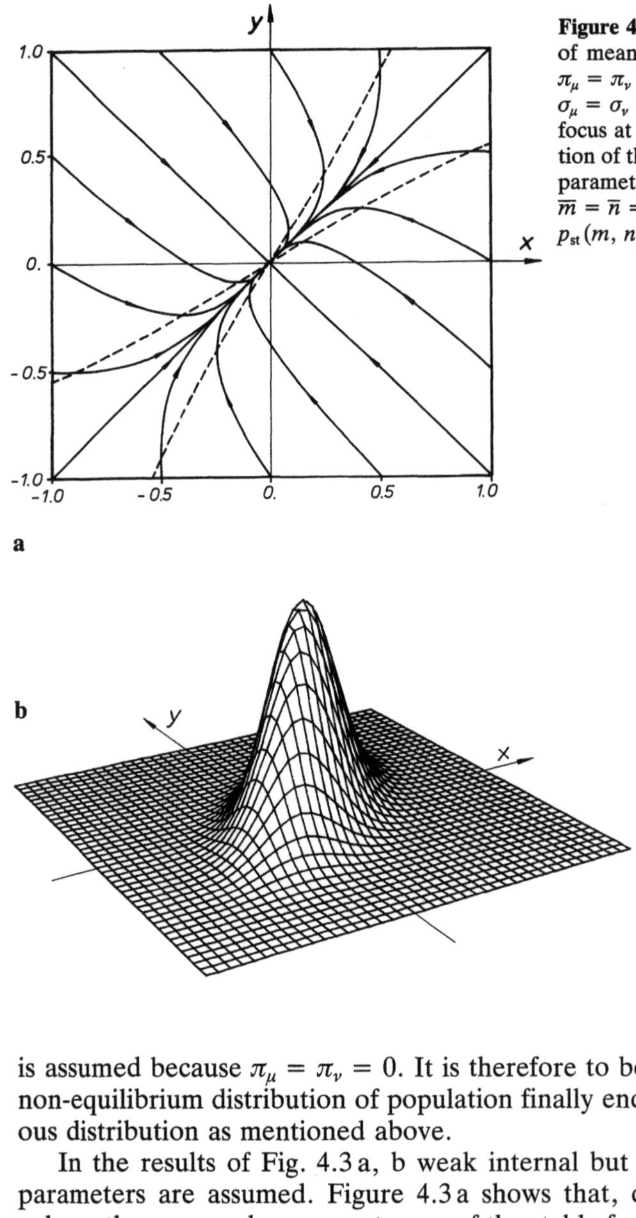

Figure 4.2. (**a**) x-y diagram of fluxlines of mean values using trend parameters $\pi_\mu = \pi_\nu = 0$, $\kappa_\mu = \kappa_\nu = 0.2$ and $\sigma_\mu = \sigma_\nu = 0.5$ leading to one stable focus at $x = y = 0$. (**b**) Stationary solution of the master equation for the trend parameters of part (a) and for $\overline{m} = \overline{n} = 20$, $p_{st}(m, n)_{max} = 0.9937 \cdot 10^{-2}$

is assumed because $\pi_\mu = \pi_\nu = 0$. It is therefore to be expected that any initial non-equilibrium distribution of population finally ends in a state of homogeneous distribution as mentioned above.

In the results of Fig. 4.3 a, b weak internal but strong external sympathy parameters are assumed. Figure 4.3 a shows that, depending on their initial values, the mean values move to one of the stable foci occuring in $\mathcal{D}_{++}$ or $\mathcal{D}_{--}$. The two foci represent a concentration of *both* populations in area 1 or area 2 of the city. On the other hand, the focus $P(0, 0)$, which represents a homogeneous population distribution, has become unstable. Thus the stationary distribution of Fig. 4.3 b shows symmetrical concentrations around the stable foci.

The following interpretation is certainly plausible: The strong external sympathy parameters imply a mutual "attraction" between the populations, in the sense of implying a trend towards a willingness to live together and therefore to migrate preferably to the area where the majority of the other population lives.

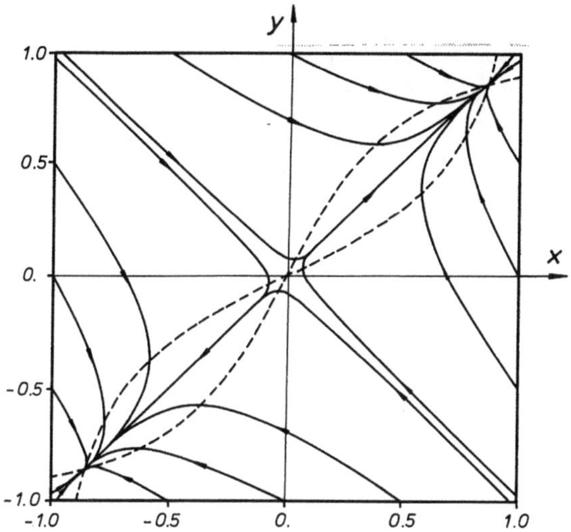

a

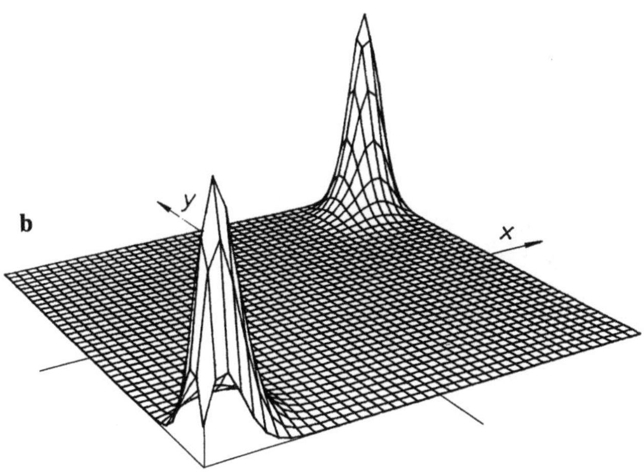

b

Figure 4.3. (**a**) x-y diagram of fluxlines of mean values using trend parameters $\pi_\mu = \pi_\nu = 0$, $\kappa_\mu = \kappa_\nu = 0.5$ and $\sigma_\mu = \sigma_\nu = 1.0$ leading to two stable foci and one unstable focus. (**b**) Stationary solution of the master equation for the trend parameters of part (a) and for $\overline{m} = \overline{n} = 20$, $p_{st}(m, n)_{max} = 0.2615 \cdot 10^{-1}$

Therefore a final and stable state is reached in which the majority of both populations settle in the *same* area of the city.

The case described by Fig. 4.3 a, b is a good example for illustrating the implication of the invariance of equation (4.53) under the transformation b: $\{x \to x;\ y \to -y;\ \sigma_\mu \to -\sigma_\mu$ and $\sigma_\nu \to -\sigma_\nu\}$ introduced above, which holds for $\pi_\nu = \pi_\mu = 0$, i.e. which holds good for all the Figures 4.2–8 with the exception of Fig. 4.5 a, b. The reflection of the figures with respect to the x-axis, i.e. $y \to -y$ implied by the transformation, is left to the imagination of the reader. The simultaneous transition $\sigma_\mu \to -\sigma_\mu$ and $\sigma_\nu \to -\sigma_\nu$ leads to new and *sociologically opposite* cases to those presented graphically, since the sympathy parameters σ_μ and σ_ν of the solutions to (4.53) belonging to the *reflected* Figures have changed their sign. In the case of Fig. 4.3 a, b the reflected configurations arise

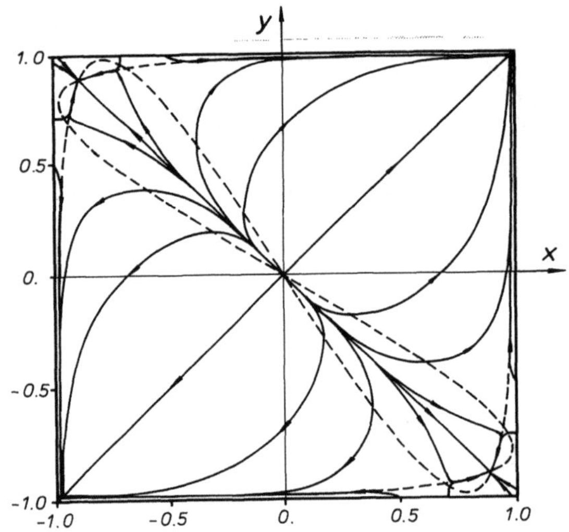

a

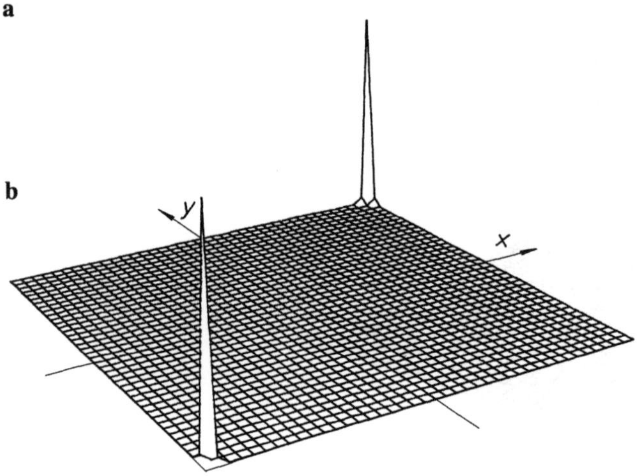

b

Figure 4.4. (**a**) x-y diagram of fluxlines of mean values using trend parameters $\pi_\mu = \pi_\nu = 0$, $\kappa_\mu = \kappa_\nu = 2.6$ and $\sigma_\mu = \sigma_\nu = 1.0$ leading to five unstable and four stable (two main and two secondary) foci. (**b**) Stationary solution of the master equation for the trend parameters of part (a) and for $\overline{m} = \overline{n} = 20$, $p_{st}(m, n)_{max} = 0.4669$

from the negative sympathy parameters $\sigma_\mu = -1$ and $\sigma_\nu = -1$. Such parameters describe a situation of mutual "antipathy" between the populations with the effect that members of the one population preferably migrate away from the area where the majority of the other population lives. The reflections of Fig. 4.3 a, b would represent the result of this change in attitudes: The fluxlines then move to the new (reflected) stable foci in the domains $\mathcal{D}_{-+}$ and $\mathcal{D}_{+-}$. The new foci represent a *separation* of the populations with either $\mathcal{P}_\mu$ living predominantly in city area 2 and $\mathcal{P}_\nu$ in area 1, or vice versa. This could be a plausible model of ghetto formation. The reader is invited to interpret for himself the reflected figures for all the above mentioned cases (where the invariance holds) along the same lines of argumentation.

For the solutions shown in Fig. 4.4 a, b extremely strong internal and strong external sympathy parameters have been assumed. Figure 4.4 a indicates that in

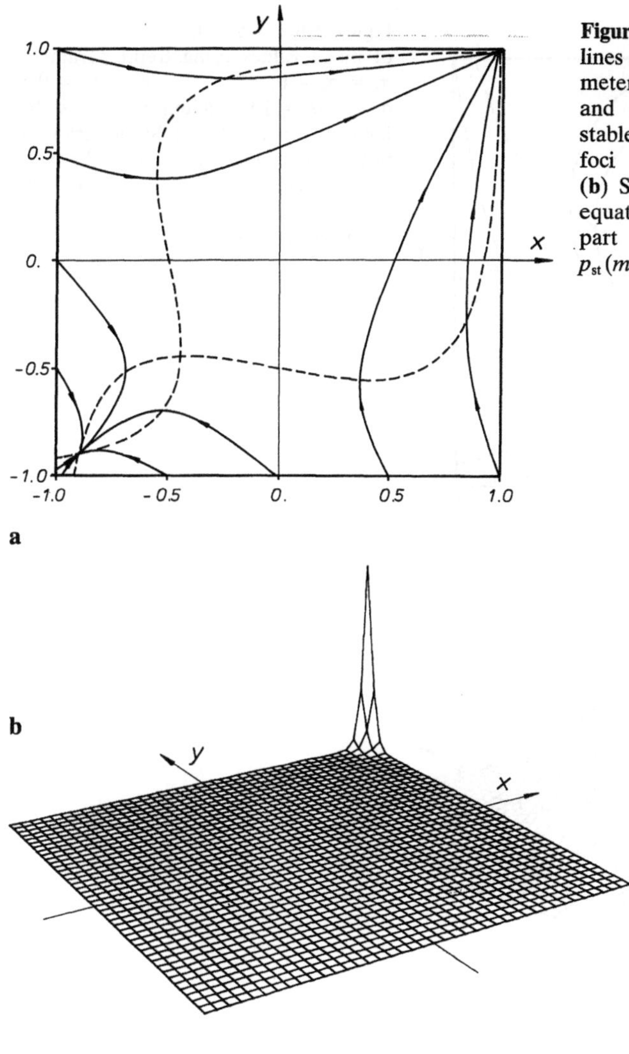

Figure 4.5. (**a**) *x-y* diagram of flux-lines of mean values using trend parameters $\pi_\mu = \pi_\nu = 0.25$, $\kappa_\mu = \kappa_\nu = 1.2$ and $\sigma_\mu = \sigma_\nu = 1.0$ leading to two stable (one main and one secondary) foci and one unstable focus. (**b**) Stationary solution of the master equation for the trend parameters of part (a) and for $\overline{m} = \overline{n} = 20$, $p_{st}(m, n)_{max} = 0.5197$

addition to the main stable foci in $\mathscr{D}_{++}$ and $\mathscr{D}_{--}$ now there exist secondary stable foci in $\mathscr{D}_{-+}$ and $\mathscr{D}_{+-}$. These correspond to a concentration of the populations $\mathscr{P}_\mu$ and $\mathscr{P}_\nu$ in *different* parts of the city in spite of the existence of a strong "mutual sympathy". This arises due to the very strong *internal* sympathy parameters prohibiting the disintegration of existing clusters of the populations $\mathscr{P}_\mu$ and $\mathscr{P}_\nu$. Depending on their initial value the mean values move to one of the main or secondary foci. The corresponding stationary distribution shown in Fig. 4.4 b, however, exhibits two pronounced maxima around the two main foci only.

Figure 4.5 a, b show the situation occuring if both populations have a preference for the city area 1, in addition to positive internal and external sympathy parameters. Fig. 4.5 a shows that the unstable and the two stable foci of Fig. 4.4 are now shifted. The mean value fluxlines now end prevailingly in the main stable focus in $\mathscr{D}_{++}$. The stationary distribution shown in Fig. 4.5 b has one

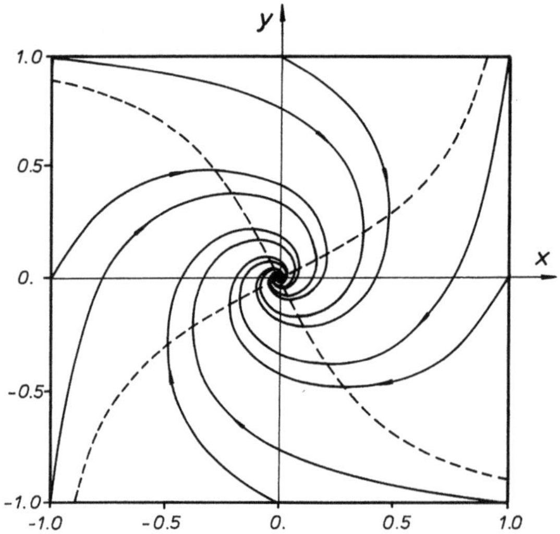

a

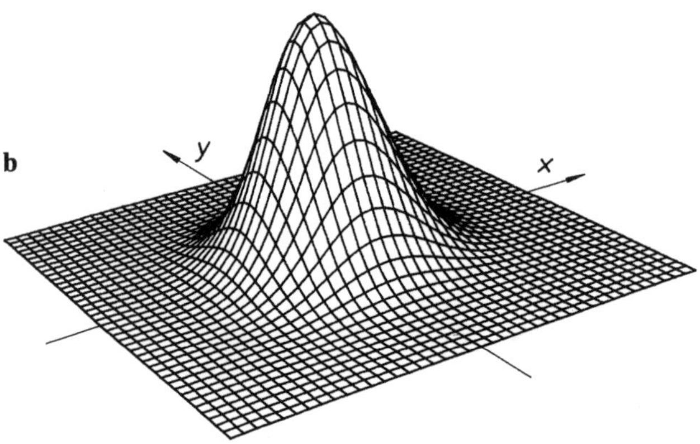

b

Figure 4.6. (**a**) x-y diagram of fluxlines of mean values using trend parameter $\pi_\mu = \pi_\nu = 0$, $\kappa_\mu = \kappa_\nu = 0.5$, $\sigma_\mu = -\sigma_\nu = 1.0$ leading to one stable focus at $x = y = 0$. (**b**) Stationary solution of the master equation for the trend parameters of part (a) and for $\overline{m} = \overline{n} = 20$; $p_{\text{st}}(m, n)_{\max} = 0.7952 \cdot 10^{-2}$

pronounced maximum only, namely around the main stable focus. This result has a clear interpretation: If the sympathy within and between the populations is superimposed by a "natual" or implicit preference for area 1, the majority of both populations will finally migrate to and settle in that area. The stable focus in $\mathcal{D}_{--}$ represents the much less probable case that – because of the existence of internal and mutual sympathy and because of the given initial situation – both populations *remain settled* in the less preferred area 2.

For the results of Fig. 4.6 a, b weak internal sympathy parameters but strong and asymmetrical external sympathy parameters have been assumed: While $\mathcal{P}_\mu$ tends to live with $\mathcal{P}_\nu$ as neighbours ($\sigma_\mu > 0$), $\mathcal{P}_\nu$ wants to separate from $\mathcal{P}_\mu$, ($\sigma_\nu < 0$). The fluxlines of Fig. 4.6 a then approach the stable focus $P(0, 0)$ in spirals expressing a new type of migration. The stationary distribution shown in Fig. 4.6 b also concentrates around this stable focus. Although the stationary

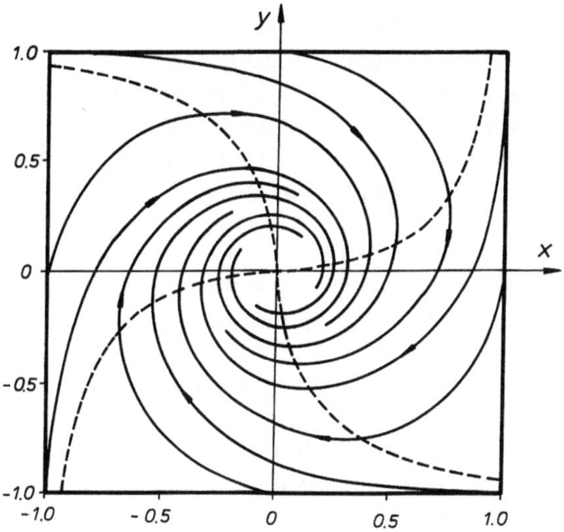

a

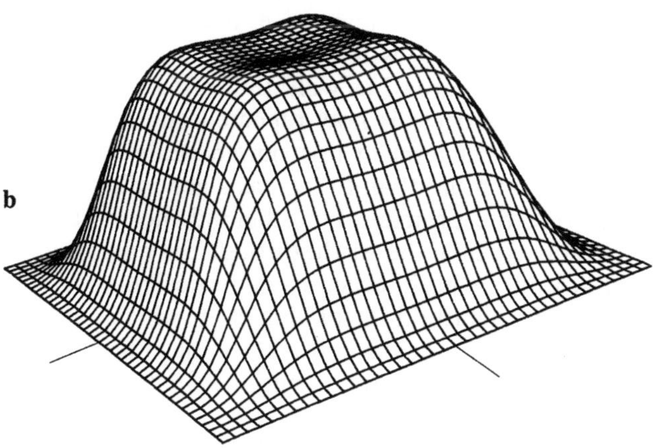

b

Figure 4.7. (**a**) x-y diagram of fluxlines of mean values using trend parameter $\pi_\mu = \pi_\nu = 0$, $\kappa_\mu = \kappa_\nu = 1.0$ and $\sigma_\mu = -\sigma_\nu = 1.0$ at the phase transition where the only focus at $x = y = 0$ becomes unstable. (**b**) Stationary solution of the master equation for the trend parameters of part (a) and for $\bar{m} = \bar{n} = 20$, $p_{st}(m, n)_{max} = 0.1323 \cdot 10^{-2}$

situation – the homogeneous distribution of both populations over both areas of the city – is the same as for the case illustrated by Fig. 4.2 a, b, the trends behind the migration process in non-equilibrium situations differ from those of Fig. 4.2 a, b since population $\mathcal{P}_\nu$ tends to avoid close contact with $\mathcal{P}_\mu$. However, the weak internal sympathy parameter associated with $\mathcal{P}_\nu$ – which indicates only a small preference of this population to live together – finally wipes out the difference between this case and that of Fig. 4.2, leading again to the stationary situation of populations distributed homogeneously over both areas.

In Fig. 4.7 a, b strong asymmetrical external sympathy parameters and internal sympathy parameters are shown to lead to a phase transition. Figure 4.7 a shows mean value fluxlines which approach the focus at $x = y = 0$ extremely slowly ("critical slowing down"). This is due to the fact that the focus becomes unstable for the values assumed for the trend parameters. The stationary distribution shown in Fig. 4.7b is now of large width around the focus corresponding to the large "critical fluctuations" of the Langevin paths at the phase transition. The interpretation of the transition from Figs. 4.6 to 4.7 is intuitively clear. Because of the choice of stronger internal sympathy parameters $\kappa_\mu = \kappa_\nu = 1.0$ instead of $\kappa_\nu = \kappa_\mu = 0.5$ the solidarity of "staying together" has been increased for each population. This favours a collective migration of groups of members of $\mathcal{P}_\mu$ and $\mathcal{P}_\nu$ separately. Correspondingly, extremely large fluctuations around the homogeneous distribution are observed (Fig. 4.7b).

The assumption of very strong internal sympathy parameters, together with the same strong asymmetrical external sympathy parameters as assumed for the results of Figs. 4.6, 7, leads to the configurations of Fig. 4.8a, b.

It can be seen in Fig. 4.8a that the mean value fluxlines now approach a limit cycle instead of a stable focus. This means that the migration process does not reach a stable situation in spite of the existence of a stationary solution to the master equation (Fig. 4.8b). This type of migration has to be interpreted as the expression of the following conflict between the populations: Starting from a configuration with both populations concentrated in city area 1 (i.e., $P(\tau_0) \in \mathcal{D}_{++}$) the population $\mathcal{P}_\nu$ – because $\sigma_\nu < 0$ – tends to avoid $\mathcal{P}_\mu$ and migrate away into area 2 and concentrate there, leaving $\mathcal{P}_\mu$ behind. This would correspond to a configuration $P(\tau_1) \in \mathcal{D}_{+-}$. $\mathcal{P}_\mu$ will follow sooner or later, however, (because $\sigma_\mu > 0$) and concentrate in area 2 as well; the configuration now has reached that of $P(\tau_3) \in \mathcal{D}_{--}$. The process is now repeated and leads both populations back to area 1 of the city, at least in this simple model. In more realistic models with a city split up into more than two areas a generalization of this type of migration may describe the principle lying behind the sequential erosion to be observed in certain big cities of the world through the migration of populations of different social attitudes who interact asymmetrically.

Figure 4.8b shows the stationary distribution for this case, which concentrates in the vicinity of the limit cycle. The peaks and connecting ridges of this distribution correspond to the varying speed of traversing the fluxlines around the limit cycle.

To conclude the discussion of the results shown in Figs. 4.2–8 a general remark on the approximate character of (4.53, 31) is appropriate. The mean value equations (4.53) used for calculating the results of Figs. 4.2a–8a will, in general, lose their validity for long periods of time. This follows already from the fact that the exact and unique mean values belonging to the stationary distributions are $x(\infty) = 0$ and $y(\infty) = 0$ in all the cases discussed and illustrated in Figs. 4.2–8 with the exception of Fig. 4.5. But the fluxlines of the approximate equation (4.53) end as $\tau \to \infty$ in one or another stable focus depending on the initial condition, or approach a limit cycle. These fluxlines approximate the path of the exact mean value only as long as the accompanying distribution is essentially uni-modal, as has already been mentioned and assumed for all the reviewed results.

In order to estimate how long a uni-modal distribution can survive in the migration process the approximate mean value and variance equations (4.53, 31) are solved, as an example, for parameters chosen as for the results of Fig. 4.8a, b and for a reasonably chosen initial variance.

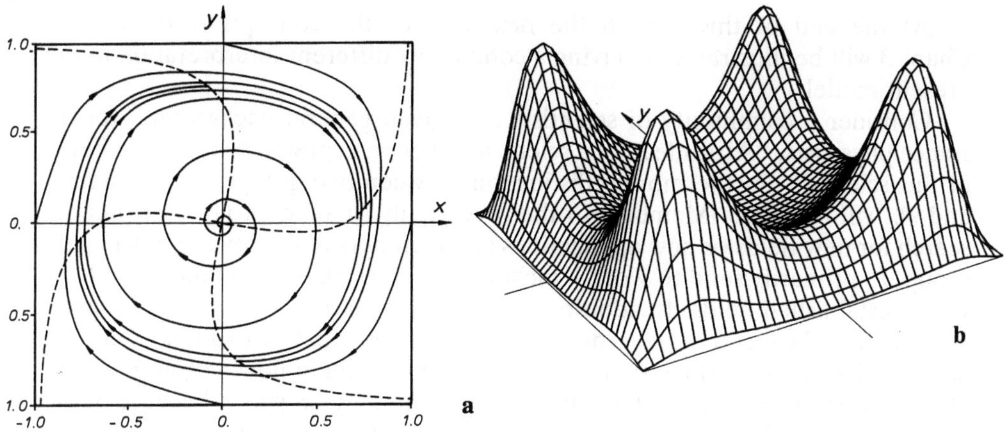

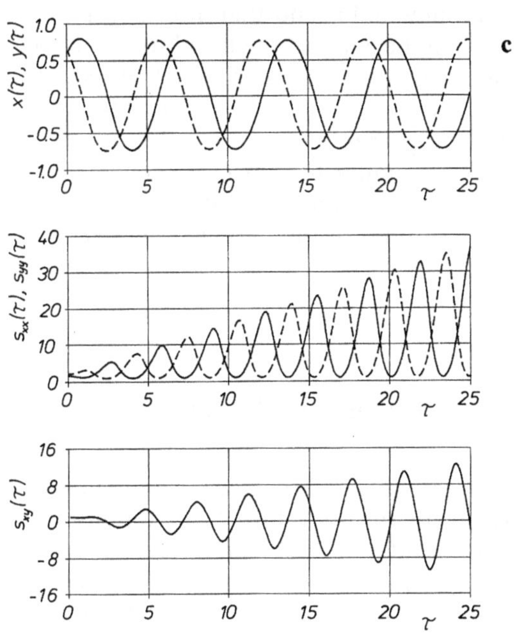

Figure 4.8. (**a**) *x*-*y* diagram of fluxlines of mean values approaching a limit cycle, using trend parameters $\pi_\mu = \pi_\nu = 0$, $\kappa_\mu = \kappa_\nu = 1.2$ and $\sigma_\mu = -\sigma_\nu = 1.0$ showing one unstable focus only at $x = y = 0$. (**b**) Stationary solution of the master equation for the trend parameters of part (a)and for $\overline{m} = \overline{n} = 20$, $p_{\text{st}}(m, n)_{\max} = 0.1957 \cdot 10^{-2}$. (**c**) Mean values $x(\tau)$ (—), $y(\tau)$ (---) and variances $s_{xx}(\tau) = \overline{m}\,\sigma_{xx}(\tau)$ (—), $s_{yy}(\tau) = \overline{m}\,\sigma_{yy}(\tau)$ (---), and $s_{xy}(\tau) = \overline{m}\,\sigma_{xy}(\tau)$ (—) as functions of scaled time $\tau = 2\gamma t$ according to equations (4.53) and (4.31) for the trend parameters $\pi_\mu = \pi_\nu = 0$, $\kappa_\mu = \kappa_\nu = 1.2$ and $\sigma_\mu = -\sigma_\nu = 1.0$ and for $\overline{m} = \overline{n} = 20$

Figure 4.8c shows that the functions $s_{xx}(\tau) = \overline{m}\,\sigma_{xx}(\tau)$, $s_{xy}(\tau) = \overline{m}\,\sigma_{xy}(\tau)$ and $s_{yy}(\tau) = \overline{m}\,\sigma_{yy}(\tau)$, whose equations of motion are independent of the value of $\overline{m} = \overline{n}$, see (4.31), consist of oscillations superimposed by a steady increase of the amplitude as the mean values $x(\tau)$ and $y(\tau)$ traverse the limit cycle. For $\overline{m} \gg 1$ and reasonable initial conditions, $|\sigma_{xx}(\tau)| \ll 1$, $|\sigma_{xy}(\tau)| \ll 1$ and $|\sigma_{yy}(\tau)| \ll 1$ may remain valid for several cycles. This implies that the distribution remains essentially uni-modal and that (4.53, 31) still hold good. Large $\sigma_{xx}(\tau)$ and small $\sigma_{yy}(\tau)$ are obtained for $x(\tau) \approx 0$ and large $|y(\tau)|$; also small $\sigma_{xx}(\tau)$ and large $\sigma_{yy}(\tau)$ are found for large $|x(\tau)|$ and $y(\tau) \approx 0$. Thus the shape of the – still uni-modal – distribution shrinks and widens alternatingly in the x and y directions during its motion around the limit cycle.

At the end of this section the flexibility of the concepts introduced in Chap. 3 will be illustrated by giving a completely different interpretation to the present model.

Consider a community of scientists of a given general background, for instance physicists. "Attitude i" of a scientist is then interpreted "working in field i of this science". Furthermore a distinction between two groups of scientists $\mathcal{P}_\nu$ and $\mathcal{P}_\mu$ will be made: Members of the (eventually smaller) group $\mathcal{P}_\nu$ primarily search for new fields or new applications and members of the (eventually larger) group $\mathcal{P}_\mu$ primarily tend to work – including team work – on detailed problems within established fields. In addition, mutations from $\mathcal{P}_\mu$ to $\mathcal{P}_\nu$ or vice versa are imaginable. The trend of the groups for change from a field 1 to a field 2 may be characterized by equations similar to (4.34, 35) with the following trend parameters – assuming $\pi_\mu = \pi_\nu = 0$ for the sake of simplicity: κ_ν small, because of the strong individualism shown by the members of $\mathcal{P}_\nu$. σ_ν negative, because the members of $\mathcal{P}_\nu$ have a wish to investigate new fields and therefore to separate themselves from the main group $\mathcal{P}_\mu$. κ_μ large, because of the trend for members of $\mathcal{P}_\mu$ to go into details and to research in teams. σ_μ positive, because the members of $\mathcal{P}_\mu$ have the attitude and therefore the wish to follow $\mathcal{P}_\nu$ into newly opened fields in order to work out details together.
This choice of parameters leads to a motion similar to, but more general than that shown in Fig. 4.8 a and describes the never ending generation and occupation of new fields of science as far as the attitudes of the participating scientists are concerned.

4.3 Birth-Death Processes Within a Single Population

In this section pure birth-death processes will be discussed. Although such models have been dealt with extensively in the literature (see, for instance [4.1–8]), there still exist some seldom considered problems referring to the relation between the exact stationary or quasi-stationary solution of the master equation and the deterministic approach. This relation will be treated in Sect. 4.3.1 including an application of the results obtained to observations on animal populations [4.15, 17]. Further, the generalization of the stochastic standard model by including multistep birth or death processes will be investigated in Sect. 4.3.2.

4.3.1 Stochastic Versus Deterministic Description

The well known model of the development of a population under birth-death processes is another special case of the general master equation (4.6) arising by considering the term $(\partial p/\partial t)_{BD}$ $(\boldsymbol{n}; t)$ only and by neglecting the migration and Volterra-Lotka terms. For *one* population and by adapting the notation to the case under consideration, the following equations are obtained:

$$\left.\begin{aligned}
\frac{dp(0;t)}{dt} &= \mu_1 p(1;t) \\
\frac{dp(1;t)}{dt} &= -(\lambda_1+\mu_1)p(1;t)+\mu_2 p(2;t) \\
&\ldots\ldots\ldots\ldots\ldots\ldots \\
\frac{dp(n;t)}{dt} &= \lambda_{n-1}p(n-1;t)-(\lambda_n+\mu_n)p(n;t)+\mu_{n+1}p(n+1;t)
\end{aligned}\right\} \quad (4.63)$$

where the birth rate $\lambda_n \equiv w(n+1, n)$ and the death rate $\mu_n \equiv w(n-1, n)$ are assumed to have the following n-dependence:

$$\left.\begin{aligned}
&\lambda_n = a_1 n \quad \text{and} \quad \mu_n = a_2 n + b_2 n^2 \quad \text{with} \\
&a_1 > 0,\ a_2 > 0 \quad \text{and} \quad b_2 > 0
\end{aligned}\right\} \quad (4.64\,\text{a})$$

or, equivalently,

$$\left.\begin{aligned}
&\lambda_n = \lambda n \quad \text{and} \quad \mu_n = \mu n + \frac{(\lambda-\mu)}{K} n^2 \quad \text{with} \\
&\lambda > \mu,\ \lambda > 0,\ \mu > 0 \quad \text{and} \quad K > 0\,.
\end{aligned}\right\} \quad (4.64\,\text{b})$$

The increase in the death rate per individual μ_n/n with growing n could, for example, be due to limitations in food resources. An analogous effect, namely a decrease in the specific birth rate λ_n/n for growing n will be neglected as it leads to essentially the same type of results. The birth and death rates per individual as defined in (4.64 b) are shown in Fig. 4.9.

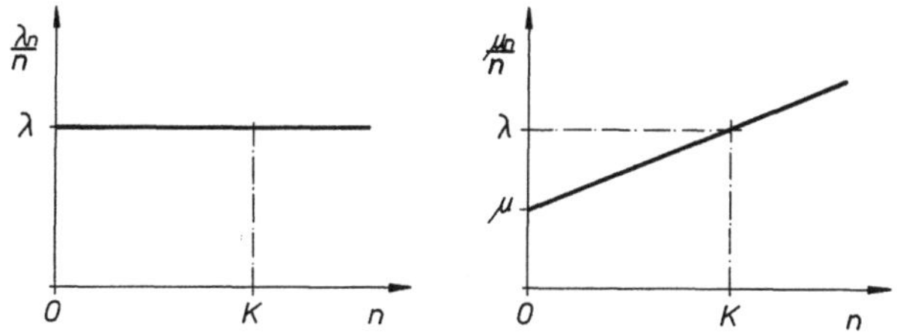

Figure 4.9. Dependence of specific birth rate λ_n/n and death rate μ_n/n on the magnitude n of the population

With these equations the condition of detailed balance is fulfilled in a trivial manner. The *exact* stationary solution, see (4.39), is obtained in the form

$$p_{st}(n) = p_{st}(0)\prod_{k=1}^{n}\xi(k) = p_{st}(0)\,\delta_{n0} \qquad (4.65)$$

where

$$\xi(n) = \frac{\lambda_{n-1}}{\mu_n} = \frac{n-1}{n} \frac{a_1}{a_2 + b_2 n} \,. \tag{4.66}$$

This solution describes a *population which has completely died out*. The solution is stable because "from nothing comes nothing". Besides this trivial solution, however, there exists a *quasi-stationary solution* for which the eventual extinction occurs extremely slowly.

In the *conventional approximation* the quasi-stationary solution is determined by neglecting $(n-1)/n$ in (4.66) and by inserting the approximate factors

$$\xi_{as}(n) \approx \frac{\lambda_n}{\mu_n} = \frac{a_1}{a_2 + b_2 n} \tag{4.67}$$

in (4.65). In this way the form

$$\left.\begin{aligned} &p_{as}(n) = p_{as}(0) \prod_{k=1}^{n} \xi_{as}(k) = p_{as}(0) \exp[-\Psi(n)] \\ &\text{with} \\ &\Psi(n) = -\sum_{k=1}^{n} \ln[\xi_{as}(k)] \approx -\int_0^n dk \ln[\xi_{as}(k)] \end{aligned}\right\} \tag{4.68}$$

for the quasi-stationary solution to (4.63) is obtained, which is asymptotically correct for large n.

The maximum $\bar{n}$ of $p_{as}(n)$ is determined by

$$\left.\frac{\partial \Psi(n)}{\partial n}\right|_{n=\bar{n}} = -\ln[\xi_{as}(\bar{n})] = 0 \tag{4.69 a}$$

or

$$\xi_{as}(\bar{n}) = 1 \qquad \text{from which} \quad \bar{n} = (a_1 - a_2)/b_2 \,. \tag{4.69 b}$$

Approximating $p_{as}(n)$ by an expansion of $\Psi(n)$ around $n = \bar{n}$:

$$\left.\begin{aligned} p_{as}(n) &= p_{as}(0) \exp\left\{-\left[\Psi(\bar{n}) + \frac{1}{2}\frac{\partial^2 \Psi(\bar{n})}{\partial \bar{n}^2}(n-\bar{n})^2 + \ldots\right]\right\} \\ &= \frac{C}{\sqrt{2\pi\sigma}} \exp\left\{-\frac{(n-\bar{n})^2}{2\sigma}\right\} \qquad \text{for} \quad n \geq 0 \\ p_{as}(n) &= 0 \qquad\qquad\qquad\qquad\qquad\qquad \text{for} \quad n < 0 \end{aligned}\right\} \tag{4.70}$$

the standard deviation σ of the distribution can be obtained as [4.3]:

$$\sigma = \left[\frac{\partial^2 \Psi(n)}{\partial n^2}\right]^{-1}_{n=\bar{n}} = -\left[\frac{\partial \xi_{as}(n)}{\partial n}\right]^{-1}_{n=\bar{n}} = \frac{a_1}{b_2} = \frac{\lambda}{(\lambda - \mu)}\bar{n} \,. \tag{4.71}$$

This conventional quasi-stationary distribution (4.68) or (4.70), however, has completely the wrong form for *small* n due to neglecting the factors $(n-1)/n$. On the other hand, it will be seen that the probability of complete extinction in the quasi-stationary solution depends decisively on the exact form for *small* values of n.

The quasi-stationary solution to (4.63) is therefore determined in a better approximation than through (4.67) by using

$$\left.\begin{aligned} p_{qs}(n) &= 0 && \text{for} \quad n = 0 \\ p_{qs}(n) &= p_{qs}(1)\prod_{k=2}^{n}\xi(k) && \text{for} \quad n \geq 1 \end{aligned}\right\} \tag{4.72}$$

together with the *exact* factors

$$\xi(n) = \frac{n-1}{n}\,\frac{a_1^*}{a_2^* + b_2^* n} \,. \tag{4.73}$$

Here, for the sake of comparing $p_{qs}(n)$ with $p_{as}(n)$, rescaled rate constants

$$\left.\begin{aligned} & a_1^* = \frac{\bar{n}}{\bar{n}-1}a_1\,, \quad a_2^* = a_2 \quad \text{and} \quad b_2^* = b_2 \\ & \text{or} \\ & \lambda^* = \frac{K}{K-1}\lambda > \lambda,\ \mu^* = \mu \quad \text{and} \quad K^* = \frac{K}{K-1}\left(K + \frac{\mu}{\lambda-\mu}\right) \end{aligned}\right\} \tag{4.74}$$

are used in order to obtain the maximum of $p_{as}(n)$ and $p_{qs}(n)$ at the same point $n = \bar{n}$. If $p_{qs}(n)$ is normalized as

$$\sum_{n=0}^{\infty} p_{qs}(n) = 1$$

the value of $p_{qs}(1)$ is obtained from (4.72) as

$$p_{qs}(1) = \left[1 + \sum_{n=2}^{\infty}\prod_{k=2}^{n}\xi(k)\right]^{-1} . \tag{4.75}$$

The turning values of the distribution $p_{qs}(n)$ are again determined by

$$\xi(n) = 1 \tag{4.76}$$

which has two positive solutions

$$n_{\pm} = \frac{K^*}{2}\left(1 \pm \sqrt{1 - 4\frac{\lambda^*}{\lambda^* - \mu^*} \cdot \frac{1}{K^*}}\right)$$

$$\text{for} \quad K^* > 4\frac{\lambda^*}{\lambda^* - \mu^*}\,. \tag{4.77}$$

Assuming $K^* \gg 4\dfrac{\lambda^*}{\lambda^* - \mu^*} > 1$ and expanding the rooted term

$$n_+ \approx K^* - \frac{\lambda^*}{\lambda^* - \mu^*} \quad \text{and}$$

$$n_- \approx \frac{\lambda^*}{\lambda^* - \mu^*} \tag{4.78}$$

are obtained. It can easily be verified that n_+ corresponds to $\bar{n}$ and gives a maximum of $p_{qs}(n)$ while $n_- \equiv n^* \ll \bar{n}$ leads to a minimum value. In Fig. 4.10 the quasi-stationary solutions $p_{as}(n)$ and $p_{qs}(n)$ obtained in the conventional way and by using the improved approximatation are compared for chosen explicit values of the rate constants.

The time paths of the distributions and moments defined by

$$\langle n^l \rangle_t = \sum_{n=0}^{\infty} n^l p(n; t) \tag{4.79}$$

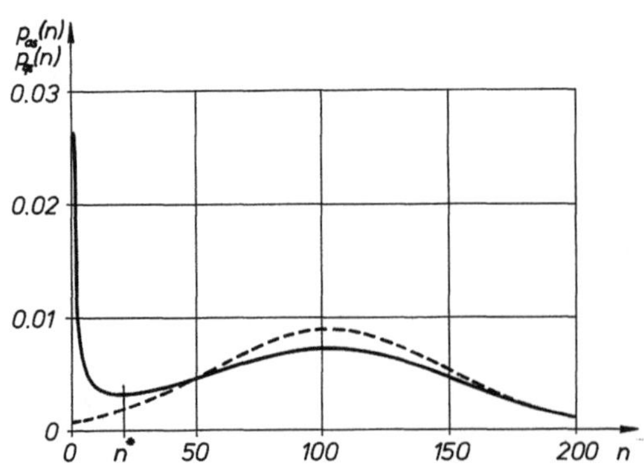

Figure 4.10. Comparison of two approximations $p_{qs}(n)$ (——) and $p_{as}(n)$ (– – –) of the quasi-stationary distribution for $a_1 = 0.5587$, $a_1^* = 0.5642$, $a_2 = 0.5320$ and $b_2 = 2.622 \cdot 10^{-4}$ with $n^* = 21$ and $\bar{n} = 102$

will now be considered. In particular, according to (4.13) equations for the mean value $\langle n\rangle_t$ and the variance $\sigma(t) = \langle n^2\rangle_t - \langle n\rangle_t^2$ are obtained as:

$$\frac{d\langle n\rangle_t}{dt} = \langle\lambda_n - \mu_n\rangle_t \tag{4.80}$$

$$\frac{d\sigma(t)}{dt} = 2(\langle n\lambda_n\rangle_t - \langle n\mu_n\rangle_t) + (1-2\langle n\rangle_t)\langle\lambda_n\rangle_t + (1+2\langle n\rangle_t)\langle\mu_n\rangle_t \tag{4.81}$$

to be solved with the initial conditions

$$\langle n\rangle_{t=0} = n_0 \quad \text{and} \quad \sigma(t=0) = \sigma_0 \, . \tag{4.82}$$

The equations (4.80, 81) are not closed. However, as long as the accompanying distribution remains essentially uni-modal, the approximations

$$\langle\lambda_n\rangle \approx \lambda_{\langle n\rangle} \quad \text{and} \quad \langle\mu_n\rangle \approx \mu_{\langle n\rangle} \tag{4.83}$$

are justified. Inserting (4.83) into (4.80, 81) and making use of (4.64b), the closed equations for $\langle n\rangle_t$ and $\sigma(t)$ are derived as

$$\frac{d\langle n\rangle_t}{dt} = r\langle n\rangle_t\left(1-\frac{\langle n\rangle_t}{K}\right) \tag{4.84}$$

$$\frac{d\sigma(t)}{dt} = (2\lambda - r)\langle n\rangle_t + \frac{r}{K}\langle n\rangle_t^2 + 2r\sigma(t)\left(1-\frac{2\langle n\rangle_t}{K}\right) \tag{4.85}$$

where $r = (\lambda - \mu)$ is the net growth rate of the population and where K is its "capacity".

Equation (4.84) is the well-known *logistic equation* of *Verhulst* and *Pearl* [4.1–5] with the solution

$$\langle n\rangle_t = K\left[1 + \frac{K-n_0}{n_0}\exp(-rt)\right]^{-1} \tag{4.86}$$

from which follows

$$\langle n\rangle_0 = n_0 \quad \text{and} \quad \langle n\rangle_\infty = K \, . \tag{4.87}$$

The equation for the variance can also be solved after inserting (4.86) into the rhs of (4.85). Furthermore, the stationary solution to (4.85) can be seen to be

$$\sigma(\infty) = \frac{\lambda}{r}K = \frac{a_1}{b_2} \tag{4.88}$$

which agrees with the variance (4.71) of the solution $p_{\mathrm{as}}(n)$.

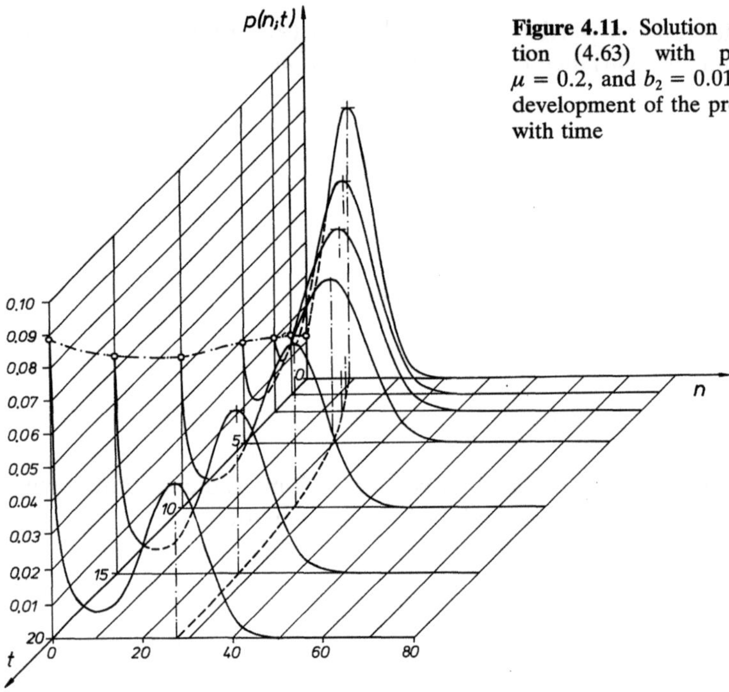

Figure 4.11. Solution of the master equation (4.63) with parameters $\lambda = 0.5$, $\mu = 0.2$, and $b_2 = 0.01$ showing the (exact) development of the probability distribution with time

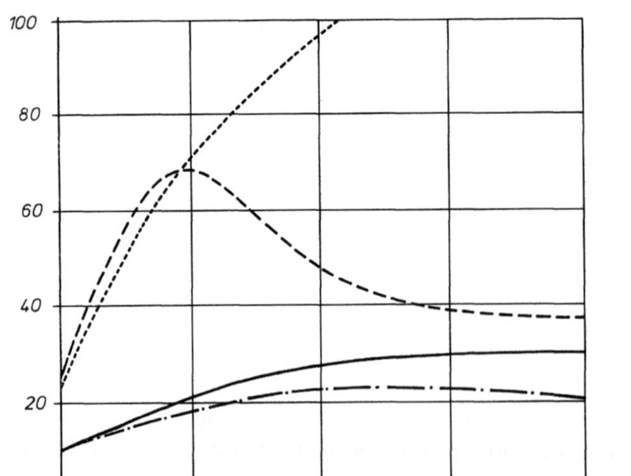

(–·–·–) $\langle n \rangle_t$, exact
(···) $\sigma(t)$, exact
(—) $\langle n \rangle_t$, solution to approximate Eq. (4.84)
(– – –) $\sigma(t)$, solution to approximate Eq. (4.85)

Figure 4.12. Time paths for the mean value $\langle n \rangle_t$ and the variance $\sigma(t)$ for the parameters $\lambda = 0.5$, $\mu = 0.2$, and $b_2 = 0.01$

The equations (4.84, 85) and their solutions are based on the approximation (4.83) which is valid for a uni-modal probability distribution only. Contrary to this assumption, however, any initially uni-modal distribution can be expected to develop first into the doubled-peaked quasi-stationary solution $p_{qs}(n)$ shown in Fig. 4.10 and then to go over finally to the exact stationary solution, that of an extinct population. Consequently deviations of the true time paths of $\langle n \rangle_t$ and $\sigma(t)$ from those described by (4.84, 85) are to be expected. A calculation of the development of a model population with time with the exact master equation (4.63) and with the parameters $\lambda = 0.5$, $\mu = 0.2$, and $b_2 = 0.01$ confirms this expectation (Figs. 4.11, 12). The Fig. 4.11 shows the exactly calculated change with time of a distribution which starts as normal distribution but soon develops into the form of the bimodal quasi-stationary distribution $p_{qs}(n)$. In Fig. 4.12 and for the same model parameters the exact paths of the mean value $\langle n \rangle_t$ and the variance $\sigma(t)$ are compared with the paths obtained by solving the approximate equations (4.84, 85).

Lastly the ideas of *Kramers* [2.5] and *van Kampen* [2.1] will be followed in calculating the extinction process and in estimating the life time of the quasi-stationary distribution. The probabilities $\pi_1(t)$ and $\pi_0(t)$ that the population is still alive or becomes extinct at time t are given by

$$\left.\begin{aligned} &\pi_1(t) = \sum_{n=1}^{\infty} p(n;t) \\ &\pi_0(t) = p(0;t) \qquad \text{with} \\ &\pi_0(t) + \pi_1(t) = 1 \,. \end{aligned}\right\} \tag{4.89}$$

From the master equation it immediately follows that

$$\frac{d\pi_0}{dt} = -\frac{d\pi_1}{dt} = \mu_1 p(1;t) \tag{4.90}$$

For $t = 0$ the quasi-stationary solution $p_{qs}(n)$ is assumed to have been established according to (4.72) with $\pi_0(0) = 0$ and $\pi_1(0) = 1$. Because the probability transfer from the living to the extinct state occurs slowly it is further assumed that the *form* of the quasi-stationary distribution is *conserved* during this process, while the *weight* of each living or non-extinct state $p(n;t)$ diminishes in proportion to $\pi_1(t)$. This implies the assumption that:

$$p(n;t) = \pi_1(t) \cdot p_{qs}(n) \qquad \text{for} \quad n = 1, 2, \ldots, \infty \,. \tag{4.91}$$

The initial state and the transfer of probability from the living to the extinct state are shown in Fig. 4.13.

Inserting (4.91) for $n = 1$ into (4.90) the equation of motion for $\pi_1(t)$ is obtained

$$\frac{d\pi_1(t)}{dt} = -\mu_1 p_{qs}(1)\,\pi_1(t) \equiv -\frac{1}{\tau}\,\pi_1(t) \tag{4.92}$$

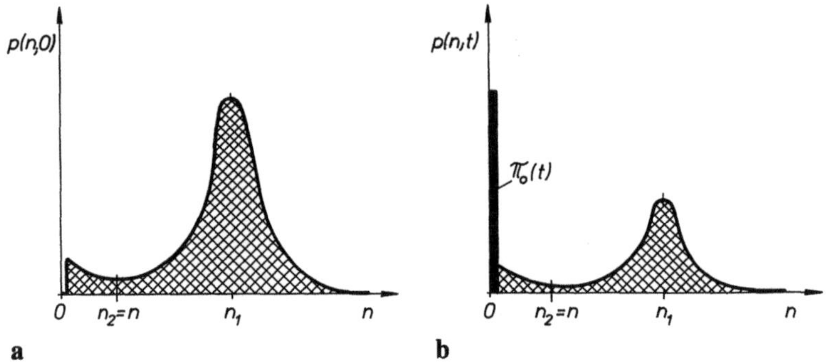

Figure 4.13. (a) Initial state of the system with $p(0; 0) = 0$ and $p(n; 0) = p_{qs}(n)$, $n = 1, 2, \ldots, \infty$, $\pi_0(0) = 0$ and $\pi_1(0) = 1$. **(b)** Transfer of probability from living states to extinct state with $p(0; t) = \pi_0(t)$ and $p(n; t) = \pi_1(t)p_{qs}(n)$, $n = 1, 2, \ldots, \infty$

with a mean life time for the quasi-stationary distribution of

$$\tau_{qs} \equiv [\mu_1 p_{qs}(1)]^{-1} = \mu_1^{-1}\left[1 + \sum_{n=2}^{\infty}\prod_{k=2}^{n}\xi(k)\right]. \tag{4.93}$$

The conventional approximation (4.68) would have yielded

$$\tau_{as} = [\mu_1 p_{as}(1)]^{-1} = \frac{\sqrt{2\pi\sigma}}{\mu_1 C}\exp\left(\frac{\bar{n}^2}{2\sigma}\right). \tag{4.94}$$

It turns out that the life time given by (4.93) may be orders of magnitude *smaller* than the life time calculated by using (4.94) because of the small but relevant mode of $p_{qs}(n)$ at $n = 1$ which is neglected in $p_{as}(n)$.

As can be seen in Fig. 4.12, there exist deviations between the exact time path of the mean value and that described by (4.84). In particular, (4.84) does not describe the extinction behaviour of $\langle n \rangle_t$. However, a modified equation for $\langle n \rangle_t$ *including* this behaviour has been proposed in the literature [4.16] with the form

$$\frac{d\langle n \rangle}{dt} = r\langle n \rangle\left(1 - \frac{\langle n \rangle}{K}\right)\frac{\langle n \rangle - n^*}{\langle n \rangle + n^*} \tag{4.95}$$

where the critical $\langle n \rangle_t$ value at $t = 0$ for extinction is $\langle n \rangle_0 = n^*$. Time paths for $\langle n \rangle_t$ according to (4.95) are shown in Fig. 4.14.

At this stage a comparison of theory with actual observations is clearly desirable. To this end, however, human populations are not particularly suitable since they underly many external influences, for example, birth control, medical aid, legislation etc., which are not described by the simple growth process modelled by (4.63) together with (4.64). It is therefore preferable to use a small quasi-stationary natural population of animals for comparison purposes, such as the population of giant Canada geese observed by *Hanson* [4.17] for several

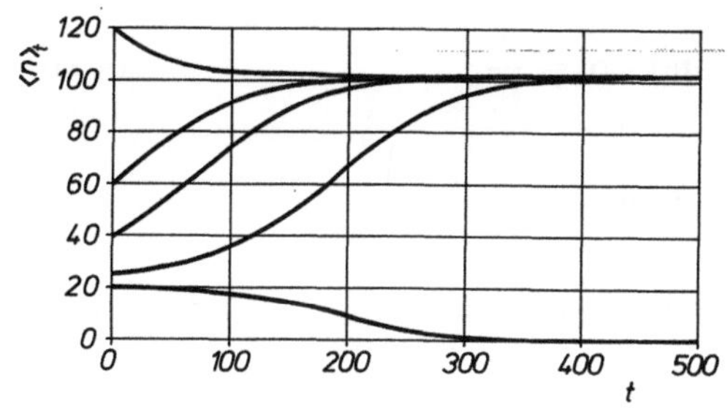

Figure 4.14. Time paths for $\langle n \rangle_t$ according to (4.95) for the parameters $r = 3.22 \cdot 10^{-2}$, $K = 102$ and $n^* = 21$

years. The observations are shown in Table 4.3 and have already been compared with theory by *De Angelis* [4.15].

Table 4.3. Spring census, gosling count and hunting harvest for giant Canada goose (Branta canadensis maxima) at Waubay National Wildlife Refuge, South Dakota, USA

Year	Total yearlings and adults in spring; n	Goslings produced, λ_n	Geese shot in vicinity of refuge ($<\mu_n$)
1945	92	58	
46	70	43	
47	130	53	
48	138	43	
49	186	91	
1950	164	110	154
51	74	70	45
52	74	46	44
53	60	33	9
54	65	60	26
1955	166	79	121
56	60	90	36
57	50	9	35
58	102	33	44
59	137	61	73
1960	89	49	48
61	115	67	22
62	50	38	29
	$\langle n \rangle = 102$, $\sigma = 2131$	$\langle \lambda_n \rangle = 57$	$\langle \mu_n \rangle_{\text{hunting}} = 52.7$

Proceeding in a somewhat different way than *De Angelis* the system parameters a_1, a_2, b_2 or λ, μ, K are determined here directly from the three measured quantities $\langle n \rangle = 102$, $\sigma = \langle n^2 \rangle - \langle n \rangle^2 = 2131$ and the birth rate $\langle \lambda_n \rangle = 57$, as these quantities seem to be the best observable ones. In addition the death rate $\langle \mu_n \rangle \approx 52.7$ is found. This value fits well to the data when it is taken into account that the total death rate should be larger than the death rate caused by hunting alone. Using the relations derived above in (4.64a, 69b, 71, 88), namely

$$\frac{\langle\lambda_n\rangle}{\langle n\rangle} = a_1 \, , \qquad \langle n\rangle = \frac{a_1 - a_2}{b_2} \quad \text{and} \quad \sigma = \frac{a_1}{b_2} \tag{4.96}$$

it is found by inserting the observations that:

$$a_1 = 0.5587, a_1^* = 0.5642, a_2 = 0.5320 \quad \text{and} \quad b_2 = 2.622 \cdot 10^{-4} \, . \tag{4.97}$$

In fact, exactly these parameters have been chosen for the comparison of the two approximations $p_{qs}(n)$ and $p_{as}(n)$ of the quasi-stationary distribution (Fig. 4.10). Using them again in (4.93, 94) to determine the mean life time of the population,

$$\tau_{qs} = 77 \text{ years} \quad \text{and} \quad \tau_{as} = 2323 \text{ years} \tag{4.98}$$

are obtained, where the value for τ_{qs} has a realistic order of magnitude, whereas the unrealistically high value for τ_{as} is due to the deficiency of the conventional approximation. The system parameters used in [4.15], however, also yield an unrealistically high value $\tau \approx 2.6 \cdot 10^7$ years which lead *De Angelis* to a criticism of the master equation approach as such. The procedure used here of determining the system parameters from the observed data and furthermore of using a better approximation of the quasi-stationary solution leads to a modification which shows that the master equation in fact *is able* to give a realistic description of small fluctuating quasi-stationary populations.

4.3.2 *Multi-Step Birth-Death Processes

So far in Sect. 4.3 only single-step birth-death processes have been discussed. But the general equations (3.15–21) can also be used to model multi-step processes (simultaneous transitions of g members of a population) and this will now be attempted here. Such multi-step processes have been investigated recently by *Haag* and *Hänggi* [4.18, 19] by using continued fraction solutions to the master equation. If the consideration here is restricted to one of the most plausible cases of multi-step birth-death processes, namely to the assumption of individual deaths but of simultaneous births of g $(g = 1, 2, \ldots, G)$ individuals [4.20], this implies that the following transition probabilities

$$\left.\begin{aligned} \mu_n &\equiv w(n-1, n) > 0, & n &= 1, 2, \ldots \\ \lambda_n^{(g)} &\equiv w(n+g, n) > 0, & n &= 1, 2, \ldots \\ & \qquad\qquad\qquad \text{with} & g &= 1, 2, \ldots, G \end{aligned}\right\} \tag{4.99}$$

are non-vanishing and that the master equation (4.63) assumes the more general form

* This section can be omitted on a first reading.

$$\left.\begin{aligned}
&\frac{dp(n;t)}{dt} = \sum_{g=1}^{G} \lambda_{n-g}^{(g)} p(n-g;t) + \mu_{n+1} p(n+1;t) - \Gamma(n) p(n;t) \\
&\text{using the abbreviation} \\
&\Gamma(n) \equiv \mu_n + \sum_{g=1}^{G} \lambda_n^{(g)} \\
&\text{for} \quad n = 0, 1, 2, \ldots, \infty \;.
\end{aligned}\right\} \tag{4.100}$$

The quasi-stationary solution to (4.100) should first be determined corresponding to the solution (4.72) for the single-step case. It is easily verified, however, that the condition of detailed balance (4.40) is *not* fulfilled in the general multi-step case and so, in fact, the quasi-stationary solution to the present problem cannot be derived in the form (4.39).

Instead of this a return to the conservation laws is made which remain valid in the multi-step case: Starting from the general form of the master equation

$$\frac{dp(i;t)}{dt} = \sum_j [w(i,j) p(j;t) - w(j,i) p(i;t)] \tag{4.101}$$

already introduced in Chap. 2, see (2.5), the states $i, j, \ldots$ can be split arbitrarily into two complementary sets A and B. Introducing probabilities for the sets of states A and B

$$\left.\begin{aligned}
p(A;t) &= \sum_{i \in A} p(i;t) \\
p(B;t) &= \sum_{j \in B} p(j;t)
\end{aligned}\right\} \tag{4.102}$$

and the probability flows between the two sets

$$\left.\begin{aligned}
j(B \leftarrow A;t) &= \sum_{j \in B} \sum_{i \in A} w(j,i) p(i;t) \\
j(A \leftarrow B;t) &= \sum_{i \in A} \sum_{j \in B} w(i,j) p(j;t)
\end{aligned}\right\} \tag{4.103}$$

from (4.101) – by taking the sum $\sum_{i \in A}$ or $\sum_{i \in B}$ on both sides – the relation

$$\begin{aligned}
\frac{dp(A;t)}{dt} &= -\frac{dp(B;t)}{dt} \\
&= j(A \leftarrow B;t) - j(B \leftarrow A;t)
\end{aligned} \tag{4.104}$$

can be derived. In the stationary case (4.104) reduces to

$$j_{st}(A \leftarrow B) = j_{st}(B \leftarrow A) \; . \tag{4.105}$$

Applying (4.105) to the master equation (4.100) and the sets $A = \{1, 2, \ldots, n-1\}$ and $B = \{n, n+1, \ldots, \infty\}$, the relation:

$$\mu_n p_{qs}(n) = \sum_{g=1}^{G} \Lambda_{n-g}^{(g)} p_{qs}(n-g) \tag{4.106}$$

with

$$\Lambda_n^{(g)} = \sum_{\nu=g}^{G} \lambda_n^{(\nu)} \tag{4.107}$$

is obtained[2]. Equation (4.106) is simpler than the stationary master equation (4.100) as it connects the probabilities of $(G+1)$ sites only instead of $(G+2)$ sites. Indeed, (4.106) can now be used to derive a continued fraction representation generalizing the representation (4.72) of $p_{qs}(n)$ to the multi-step case. Introducing the ratios

$$\eta_n = \frac{p_{qs}(n-1)}{p_{qs}(n)} \, , \qquad n = 2, 3, \ldots, \infty \tag{4.108}$$

and hence the representation

$$p_{qs}(n) = p_{qs}(1) \left[\prod_{\nu=2}^{n} \eta_\nu \right]^{-1} , \qquad n = 2, 3, \ldots, \infty \tag{4.109}$$

the following can be obtained from (4.106) after dividing by $p_{qs}(n)$

$$\mu_n = \sum_{g=1}^{G} \Lambda_{n-g}^{(g)} \prod_{\mu=1}^{g} \eta_{n+1-\mu} \tag{4.110 a}$$

or

$$\eta_n = \frac{\mu_n}{\left(\Lambda_{n-1}^{(1)} + \sum_{g=2}^{G} \Lambda_{n-g}^{(g)} \prod_{\mu=2}^{g} \eta_{n+1-\mu} \right)} \, . \tag{4.110 b}$$

[2] Setting $\mu_1 = 0$ leads to a stable distribution $p_{qs}(n)$ for $n = 1, 2, \ldots, \infty$, exactly as for the single-step master equation.

The form (4.110 b) yields a finite continued fraction representation of the η_n beginning with

$$\eta_1 = 0 \qquad \text{and} \quad \eta_2 = \frac{\mu_2}{\Lambda_1^{(1)}} \tag{4.111}$$

if $\mu_1 = 0$. Using (4.110 b), η_n (and hence $p_{qs}(n)$) can be determined recursively. For $G = 1$ (single-step processes only)

$$\eta_n = \frac{\mu_n}{\Lambda_{n-1}^{(1)}} = \frac{\mu_n}{\lambda_{n-1}^{(1)}} \tag{4.112}$$

is obtained. In this case (4.109) reduces to (4.72) with (4.66), as expected.

The general multi-step solution (4.109), now written in the form

$$\left.\begin{aligned} & p_{qs}^{(G)}(n) = p_{qs}^{(G)}(1)\exp[-\Phi(n)] \\ & \text{where} \\ & \Phi(n) = \sum_{\nu=2}^{n} \ln(\eta_\nu) \approx \int_1^n d\nu \ln[\eta(\nu)] \end{aligned}\right\} \tag{4.113}$$

can be compared with the single-step solution $p_{as}(n)$ of (4.68). For the birth and death rates an assumption is made which generalizes (4.64 a)

$$\left.\begin{aligned} \lambda_n^{(g)} &= a_1^{(g)} n\,, \qquad g = 1, 2, \dots, G \\ \mu_n &= a_2 n + b_2 n^2\,. \end{aligned}\right\} \tag{4.114}$$

The maximum of $p_{qs}^{(G)}(n)$ is then determined by analogy to (4.69 a, b) as

$$\left.\frac{\partial \Phi(n)}{\partial n}\right|_{n=\bar{n}} = \ln[\eta(\bar{n})] = 0 \tag{4.115 a}$$

or

$$\eta(\bar{n}) = 1\,. \tag{4.115 b}$$

Inserting this condition for $\bar{n}$ into the continued fraction representation (4.110 b) and assuming slowly varying η_n and $\Lambda_{\bar{n}}$, i.e. $\Lambda_{\bar{n}-g} \approx \Lambda_{\bar{n}}$ and $\eta_{\bar{n}+1-\mu} \approx \eta_{\bar{n}} = 1$,

$$1 = \frac{\mu_{\bar{n}}}{\sum_{g=1}^{G} \Lambda_{\bar{n}}^{(g)}} = \frac{\mu_{\bar{n}}}{\sum_{g=1}^{G} g\lambda_{\bar{n}}^{(g)}} \tag{4.116}$$

is obtained or, using (4.114)

$$\bar{n} = \frac{\sum_{g=1}^{G} g a_1^{(g)} - a_2}{b_2} . \tag{4.117}$$

Comparing (4.117) with (4.69 b) it is evident that it is necessary to have

$$\sum_{g=1}^{G} g a_1^{(g)} = a_1 \tag{4.118}$$

in order to obtain the maximum of the distribution at the same point $n = \bar{n}$ in both the cases of single-birth and multiple-birth processes.

The dispersion of $p_{qs}^{(G)}(n)$ is determined by analogy to (4.71) as:

$$\sigma_G = \left[\frac{\partial^2 \Phi(n)}{\partial n^2}\right]_{n=\bar{n}}^{-1} = \left[\frac{\partial \eta(n)}{\partial n}\right]_{n=\bar{n}}^{-1} \tag{4.119}$$

where the derivative of $\eta(n)$ is found by differentiating the approximate form of the recurrence relation (4.110 b)

$$\sum_{g=1}^{G} \Lambda_n^{(g)} \eta_n^g \approx \mu_n . \tag{4.120}$$

This differentiation yields for $n = \bar{n}$ where $\eta_{\bar{n}} = 1$, and using the notation $\partial_n \equiv \partial/\partial_n$

$$\partial_{\bar{n}} \eta_{\bar{n}} = \frac{\partial_{\bar{n}} \mu_{\bar{n}} - \sum_{g=1}^{G} \partial_{\bar{n}} \Lambda_{\bar{n}}^{(g)}}{\sum_{g=1}^{G} g \Lambda_{\bar{n}}^{(g)}} . \tag{4.121}$$

Inserting this result into (4.119) and using (4.107, 114, 117), the dispersion

$$\sigma_G = \frac{\sum_{g=1}^{G} g \left(\sum_{\nu=g}^{G} a_1^{(\nu)}\right)}{b_2} = \frac{\sum_{g=1}^{G} \frac{g(g+1)}{2} a_1^{(g)}}{b_2} \tag{4.122}$$

of the distribution in the multiple birth case is obtained, which can be compared with the corresponding dispersion (4.71) in the single birth case using the equivalent total birth rate of (4.118)

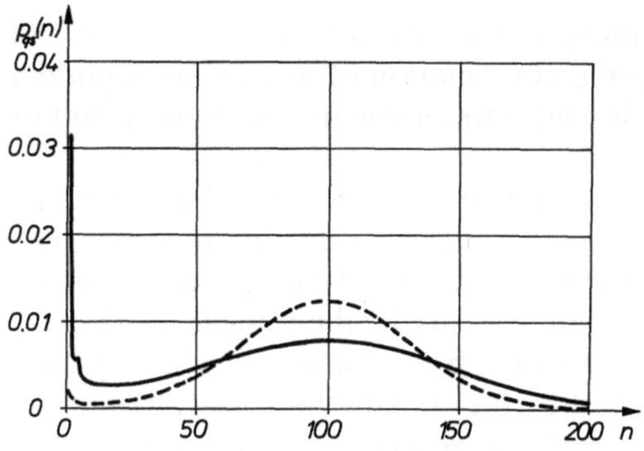

Figure 4.15. Comparison of two quasi-stationary distributions: 1) the triple birth process distribution $p_{qs}^{(3)}(n)$ (—) with parameters $a_1^{(3)} = 0.1533$, $a_1^{(g)} = 0$ for $g \neq 3$, $a_2 = 0.4$ and $b_2 = 0.5 \cdot 10^{-3}$; 2) the single birth process distribution $p_{qs}^{(1)}(n)$ (---) with parameters $a_1^{(1)} = 0.4545$, $a_1^{(g)} = 0$ for $g \neq 1$, $a_2 = 0.4$ and $b_2 = 0.5 \cdot 10^{-3}$

$$\sigma_{G=1} = \frac{a_1}{b_2} = \frac{\sum_{g=1}^{G} g a_1^{(g)}}{b_2} . \tag{4.123}$$

The ratio σ_G/σ_1 is larger than 1 and increases with the weight of the higher multiple-birth processes. This increase of the width of the distribution, for which an explicit form has now been found, is not unexpected, as multi-step processes always imply a higher fluctuation than single-step processes. This effect becomes evident in Fig. 4.15 in which two given distributions are compared:

1) a distribution $p_{qs}^{(3)}(n)$ according to (4.113), which is generated by *triple birth processes* only and whose mean value $\bar{n}$ and variance σ is chosen in agreement with $\bar{n}$ and σ of $p_{qs}(n)$ in Fig. 4.10. The triple birth distribution $p_{qs}^{(3)}(n)$ shows a new *fine structural for small n* which is not present in the (single-birth) distribution $p_{qs}(n)$ of Fig. 4.10.

2) A distribution $p_{qs}^{(1)}(n)$ generated by *single-birth processes* only with the same mean value $\bar{n}$ and birth-death rates chosen in correspondence to those of $p_{qs}^{(3)}(n)$. The distribution $p_{qs}^{(1)}(n)$ has an evidently *smaller dispersion* than $p_{qs}^{(3)}(n)$ due to the *smaller fluctuations* generated by single birth processes.

Calculating the mean life times of the quasi-stationary distributions $p_{qs}^{(1)}(n)$ and $p_{qs}^{(3)}(n)$ for the *Hanson* [4.17] observations presented at the end of Sect. 4.3.1, $\tau^{(1)} = 1460$ years and $\tau^{(3)} = 79$ years are obtained.

4.4 Migration and Predator-Prey Interaction Between Two Species

This section is meant to contribute to the old problem of the interaction between two biological species. "To be or not to be" is the essential question decided by the predator-prey interaction for the members of certain species. The famous Volterra-Lotka model for this problem [4.21, 22] has attracted many researchers who have tried to generalize it in many respects [4.1, 6, 8, 9, 23–26].

As in the preceding section it is attempted to add a new aspect to an old problem: In this case the new aspect is the combination of *non-linear migration* with the predator-prey interaction. It is useful to anticipate one main result of considering this combination:

The *pure* Volterra-Lotka model leads to a *mathematically marginal case*: The singular point (= stationary state) is a *center* encircled by *closed orbits*. The latter are non-stationary solutions *depending on the initial conditions* in analogy to the solutions of undamped hamiltonian systems in theoretical mechanics. Therefore the original model is sensitive to small perturbations which may have different causes [4.1, 23–26, 30, 31]. On the other hand, Volterra-Lotka cycles have been observed by biologists in nature for several entirely different species [4.21, 22, 27–29]. It has been stressed that the existence of such cycles could depend on the availability of appropriate facilities for migration [4.28, 31].

The extension of the pure Volterra-Lotka model by adding *non-linear migration* leads to a complex picture with a relatively large variety of different solution cases. In particular, the result has been obtained that with an appropriate "mixture" of migration and predator-prey interaction a *limit cycle* appears which is similar to the original Volterra-Lotka cycles but which is stable and *independent of the initial conditions.*

4.4.1 Master Equation and Mean Value Equations for the Special Model

The model proposed is both simple and plausible: Two habitats are assumed, an open habitat 1 and a refuge habitat 2. The prey species is able to migrate between both habitats, while for the predatory species only the open habitat 1 is accessible; this situation is illustrated in Figure 4.16. Evidently a simple socio-configuration $\{m_1, m_2, n\}$ can consist of three integers, the numbers m_1 and m_2 of animals that can be preyed upon in the habitats 1 and 2 and the number $n_1 \equiv n$ of predatory animals in, and restricted to, habitat 1. Denoting the prey and predatory populations by $\mathcal{P}_\delta$ and $\mathcal{P}_\beta$ and using the notation of Sect. 4.1, the following contributions to the configuration transition probabilities are obtained:

Migration of $\mathcal{P}_\delta$

$$\left.\begin{aligned} w_{21}^{\delta}[-1, 1, 0; m_1, m_2, n] &= m_1 \mu_{21}(m_1, m_2, n) \\ w_{12}^{\delta}[1, -1, 0; m_1, m_2, n] &= m_2 \mu_{12}(m_1, m_2, n) \end{aligned}\right\} \tag{4.124}$$

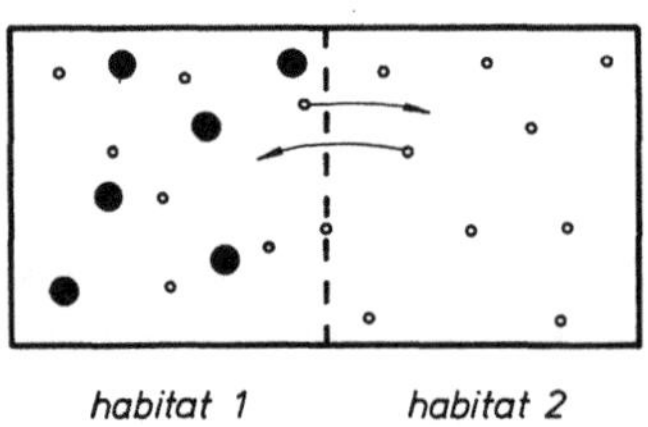

Figure 4.16. Prey (○) and predatory (●) populations in two habitats 1 and 2. Only the prey population is able to migrate between the open habitat 1 and the refuge habitat 2

where the following forms for the individual migration probabilities are adopted

$$\left.\begin{aligned} \mu_{12}(m_1, m_2, n) &= \hat{\nu}\exp[u(m_1, m_2, n)] \\ \mu_{21}(m_1, m_2, n) &= \hat{\nu}\exp[-u(m_1, m_2, n)] \\ \text{with} & \\ u(m_1, m_2, n) &= [\hat{\delta} + \hat{\alpha}(m_1 - m_2) - \hat{\beta} n] \, . \end{aligned}\right\} \quad (4.125)$$

In accordance with their substantial implications the trend parameters $\hat{\nu}$, $\hat{\delta}$, $\hat{\alpha}$ and $\hat{\beta}$ are named as follows:

$\hat{\nu}$: flexibility parameter
$\hat{\delta}$: preference parameter
$\hat{\alpha}$: clustering parameter for $\mathcal{P}_\delta$
$\hat{\beta}$: fear parameter ($\mathcal{P}_\delta$ fears $\mathcal{P}_\beta$)

Birth-death processes in $\mathcal{P}_\delta$ *and* $\mathcal{P}_\beta$

$$\left.\begin{aligned} w_{1+}^{\delta}[1, 0, 0; m_1, m_2, n] &= m_1 \beta_1^{\delta} \\ w_{1-}^{\delta}[-1, 0, 0; m_1, m_2, n] &= m_1 \delta_1^{\delta} + m_1^2 \gamma_1^{\delta} \end{aligned}\right\} \quad (4.126\,\text{a})$$

$$\left.\begin{aligned} w_{2+}^{\delta}[0, 1, 0; m_1, m_2, n] &= m_2 \beta_2^{\delta} \\ w_{2-}^{\delta}[0, -1, 0; m_1, m_2, n] &= m_2 \delta_2^{\delta} + m_2^2 \gamma_2^{\delta} \end{aligned}\right\} \quad (4.126\,\text{b})$$

$$\left.\begin{aligned} w_{1+}^{\beta}[0, 0, 1; m_1, m_2, n] &= n \beta_1^{\beta} \\ w_{1-}^{\beta}[0, 0, -1; m_1, m_2, n] &= n \delta_1^{\beta} + n^2 \gamma_1^{\beta} \, . \end{aligned}\right\} \quad (4.126\,\text{c})$$

Prey-predator interaction between $\mathcal{P}_\delta$ *and* $\mathcal{P}_\beta$

$$w_1^{\delta\beta}[-1, 0, 1; m_1, m_2, n] = m_1 n \, v_{\delta\beta} \, . \quad (4.127)$$

For convenience, the parameters are specialized and given a simplified notation for the model to be presented and numerically implemented here, thus

$$\left.\begin{aligned} &\beta_1^{\delta} = \beta_2^{\delta} = \kappa_1, \ \gamma_2^{\delta} = \omega, \ \delta_1^{\beta} = \kappa_2, \ v_{\delta\beta} = \gamma \\ &\delta_1^{\delta} = \delta_2^{\delta} = 0, \ \gamma_1^{\delta} = 0, \ \beta_1^{\beta} = 0 \qquad \text{and} \quad \gamma_1^{\beta} = 0 \, . \end{aligned}\right\} \quad (4.128)$$

The total configuration transition probability $w[k_1, k_2, k_3; m_1, m_2, n]$ is the sum of the transition probabilities (4.124, 126, 127) for the individual partial processes, see (4.5).

The master equation for the probability distribution $p(m_1, m_2, n; t)$ over the socio-configuration has now the general form of (4.6), namely

$$\frac{dp(m_1, m_2, n; t)}{dt}$$

$$= \left(\frac{\partial p}{\partial t}\right)_{\mathrm{M}}(m_1, m_2, n; t) + \left(\frac{\partial p}{\partial t}\right)_{\mathrm{BD}}(m_1, m_2, n; t) + \left(\frac{\partial p}{\partial t}\right)_{\mathrm{VL}}(m_1, m_2, n; t) \tag{4.129}$$

where the three terms on the rhs have the explicit forms

$$\left.\begin{aligned}
&\left(\frac{\partial p}{\partial t}\right)_{\mathrm{M}}(m_1, m_2, n; t)\\
&\quad = (m_1+1)\mu_{21}(m_1+1, m_2-1, n)\,p(m_1+1, m_2-1, n; t)\\
&\quad -m_1\mu_{21}(m_1, m_2, n)\,p(m_1, m_2, n; t)\\
&\quad +(m_2+1)\mu_{12}(m_1-1, m_2+1, n)\,p(m_1-1, m_2+1, n; t)\\
&\quad -m_2\mu_{12}(m_1, m_2, n)\,p(m_1, m_2, n; t)
\end{aligned}\right\} \tag{4.130}$$

$$\left.\begin{aligned}
&\left(\frac{\partial p}{\partial t}\right)_{\mathrm{BD}}(m_1, m_2, n; t)\\
&\quad = \kappa_1(m_1-1)\,p(m_1-1, m_2, n; t) - \kappa_1 m_1 p(m_1, m_2, n; t)\\
&\quad + \kappa_1(m_2-1)\,p(m_1, m_2-1, n; t) - \kappa_1 m_2 p(m_1, m_2, n; t)\\
&\quad + \omega(m_2+1)^2 p(m_1, m_2+1, n; t) - \omega m_2^2 p(m_1, m_2, n; t)\\
&\quad + \kappa_2(n+1)\,p(m_1, m_2, n+1; t) - \kappa_2 n\,p(m_1, m_2, n; t)
\end{aligned}\right\} \tag{4.131}$$

and

$$\left.\begin{aligned}
&\left(\frac{\partial p}{\partial t}\right)_{\mathrm{VL}}(m_1, m_2, n; t)\\
&\quad = \gamma(m_1+1)(n-1)\,p(m_1+1, m_2, n-1; t)\\
&\quad - \gamma m_1 n\,p(m_1, m_2, n; t)\,.
\end{aligned}\right\} \tag{4.132}$$

From the master equation (4.129) (for concentrated uni-modal distributions) and according to (4.14) the approximate closed equations for the mean values $\tilde{x} \equiv \langle m_1\rangle_t$, $\tilde{y} \equiv \langle m_2\rangle_t$ and $\tilde{z} \equiv \langle n\rangle_t$ are obtained as

$$\frac{d\tilde{x}}{dt} = -\{\tilde{x}\mu_{21}(\tilde{x}, \tilde{y}, \tilde{z}) - \tilde{y}\mu_{12}(\tilde{x}, \tilde{y}, \tilde{z})\} + \kappa_1\tilde{x} - \gamma\tilde{x}\cdot\tilde{z} \tag{4.133 a}$$

$$\frac{d\tilde{y}}{dt} = +\{\tilde{x}\mu_{21}(\tilde{x}, \tilde{y}, \tilde{z}) - \tilde{y}\mu_{12}(\tilde{x}, \tilde{y}, \tilde{z})\} + \kappa_1\tilde{y} - \omega\tilde{y}^2 \tag{4.133 b}$$

$$\frac{d\tilde{z}}{dt} = -\kappa_2\tilde{z} + \gamma\tilde{x}\tilde{z} . \tag{4.133 c}$$

After a transformation to scaled variables and scaled time

$$x = \frac{\kappa_2}{\gamma}\tilde{x} , \quad y = \frac{\kappa_2}{\gamma}\tilde{y} , \quad z = \frac{\kappa_1}{\gamma}\tilde{z} \quad \text{and} \quad \tau = \kappa_1 t \tag{4.134}$$

as well as to scaled trend parameters

$$\left.\begin{array}{l} \nu = \dfrac{\hat{\nu}}{\kappa_1} , \quad \delta = \hat{\delta} , \quad \alpha = \dfrac{\gamma}{\kappa_2}\hat{\alpha} , \quad \beta = \dfrac{\gamma}{\kappa_1}\hat{\beta} , \\[2ex] \dfrac{1}{y_0} = \dfrac{\omega}{\kappa_1} \cdot \dfrac{\kappa_2}{\gamma} \quad \text{and} \quad a = \dfrac{\kappa_2}{\kappa_1} \end{array}\right\} \tag{4.135}$$

and to

$$\hat{u} = \hat{\delta} + \hat{\alpha}(\tilde{x} - \tilde{y}) - \hat{\beta}\tilde{z} = u = \delta + \alpha(x - y) - \beta z \tag{4.136}$$

the final scaled and explicit form of the mean value equations are obtained from (4.133) making use of (4.125) as

$$\frac{dx}{d\tau} = -\nu\{x\exp(-u) - y\exp(u)\} + x(1 - z) \tag{4.137 a}$$

$$\frac{dy}{d\tau} = \nu\{x\exp(-u) - y\exp(u)\} + \left(\frac{y_0 - y}{y_0}\right) y \tag{4.137 b}$$

$$\frac{dz}{d\tau} = -a(1 - x)z . \tag{4.137 c}$$

The equations (4.137 a–c) build the starting point of the following analysis.

4.4.2 Comparison of Predator-Prey Interaction Without and With Non-Linear Migration

As far as the singular points and their stability are concerned, standard analytical procedures will be applied. Evidently, the singular points $P_s(x_s, y_s, z_s)$ of the system are solutions to the three transcendental equations:

$$0 = -\nu[x_s\exp(-u_s) - y_s\exp(u_s)] + x_s(1 - z_s) \tag{4.138 a}$$

$$0 = \nu[x_s\exp(-u_s) - y_s\exp(u_s)] + \left(\frac{y_0 - y_s}{y_0}\right) y_s \tag{4.138 b}$$

$$0 = -a(1-x_s)\,z_s\,. \tag{4.138c}$$

The linearized equations of motion (4.137) around a singular point $P_s(x_s, y_s, z_s)$ are then

$$\frac{d}{d\tau}\begin{pmatrix}\xi\\ \eta\\ \zeta\end{pmatrix} = \begin{pmatrix}a_{11} & a_{12} & a_{13}\\ a_{21} & a_{22} & a_{23}\\ a_{31} & a_{32} & a_{33}\end{pmatrix}\begin{pmatrix}\xi\\ \eta\\ \zeta\end{pmatrix} \tag{4.139}$$

where

$$\left.\begin{aligned} x(\tau) &= x_s + \xi(\tau)\\ y(\tau) &= y_s + \eta(\tau)\\ z(\tau) &= z_s + \zeta(\tau)\end{aligned}\right\} \tag{4.140}$$

and where the matrix elements in (4.139) are defined as

$$\left.\begin{aligned} a_{11} &= 1 - z_s + \nu(\alpha x_s - 1)\exp(-u_s) + \nu\alpha y_s \exp(u_s)\\ a_{12} &= -[\nu\alpha x_s \exp(-u_s) + \nu(\alpha y_s - 1)\exp(u_s)]\\ a_{13} &= -[x_s + \nu\beta x_s \exp(-u_s) + \nu\beta y_s \exp(u_s)]\\ a_{21} &= -[\nu(\alpha x_s - 1)\exp(-u_s) - \nu\alpha y_s \exp(u_s)]\\ a_{22} &= \frac{(y_0 - y_s)}{y_0} - \frac{y_s}{y_0} + \nu\alpha x_s \exp(-u_s) + \nu(\alpha y_s - 1)\exp(u_s)\\ a_{23} &= \nu\beta x_s \exp(-u_s) + \nu\beta y_s \exp(u_s)\\ a_{31} &= a z_s\\ a_{32} &= 0\\ a_{33} &= -a(1-x_s)\,.\end{aligned}\right\} \tag{4.141}$$

For the (linear) stability analysis, (4.139) is solved using

$$\begin{pmatrix}\xi(\tau)\\ \eta(\tau)\\ \zeta(\tau)\end{pmatrix} = \begin{pmatrix}\xi_0\\ \eta_0\\ \zeta_0\end{pmatrix}\exp(\lambda\tau) \tag{4.142}$$

where the eigenvalue λ has to satisfy the secular equation

$$P(\lambda) \equiv \begin{vmatrix} a_{11}-\lambda & a_{12} & a_{13}\\ a_{21} & a_{22}-\lambda & a_{23}\\ a_{31} & a_{32} & a_{33}-\lambda\end{vmatrix} = 0\,. \tag{4.143}$$

Whereby it has to be decided whether *all* the eigenvalues λ_j fulfill the condition $\mathrm{Re}\{\lambda_j\} \leq 0$ for $j = 1, 2, 3$. $P_s(x_s, y_s, z_s)$ is stable *only in this case.*

Three particular cases will be analysed further:

A) the pure Volterra-Lotka model by neglecting migration and saturation of $\mathcal{P}_\delta$ in the refuge habitat
B) the full model including migration and saturation of $\mathcal{P}_\delta$ the refuge habitat by choosing trend parameters allowing for particularly simple, but non-degenerate solutions to (4.138)
C) the model including migration but neglecting saturation of $\mathcal{P}_\delta$

Case A. For the pure Volterra-Lotka model it is necessary to set $\nu = 0$; this leads to a vanishing migration probability and decouples habitat 1 from habitat 2. Taking into account habitat 1 only, from (4.138 a, c) two singular points are found: $P_0(x = 0,\ z = 0)$ and $P_1(x = 1,\ z = 1)$. The eigenvalues of the linearized equations (4.139) obey (4.143) and are

$$\left.\begin{array}{l} \lambda_1^{(0)} = 1,\ \lambda_2^{(0)} = -a \\ \lambda_1^{(1)} = \mathrm{i}\sqrt{a},\ \lambda_2^{(1)} = -\mathrm{i}\sqrt{a} \end{array}\right\} \tag{4.144}$$

for P_0 and P_1, respectively. Hence P_0 is unstable and P_1 is a "center". The form of the orbits around the center is not only known in the vicinity of P_1, because, using the exact equations of motion (4.137 a, c), it can easily be shown that

$$[x(\tau) - \ln x(\tau)] + \frac{1}{a}[z(\tau) - \ln z(\tau)] = K \tag{4.145}$$

is constant, i.e., the lhs is a constant of motion and (4.145) determines the shape of all orbits in the x-z plane. It has already been mentioned that a center is a marginal case of a singular point and that the addition of small perturbations in the equations of motion can change the nature of the solutions of the pure Volterra-Lotka model.

Case B. The full model with migration of the prey and saturation in the refuge habitat will now be considered.

The equations (4.138 a–c) have a trivial singular point solution $P_0(0, 0, 0)$ which, considering (4.137), turns out to be unstable

$$\lambda_1 = -a,\ \lambda_2 = 1,\ \lambda_3 = 1 - 2\nu\cosh(\delta)\ .$$

For the non-trivial solutions it follows from (4.138 c) that

$$\begin{array}{lll} \text{either} & \text{I)} & x_s = 1 \\ \text{or} & \text{II)} & z_s = 0\ . \end{array} \tag{4.146}$$

For I) it follows from (4.138 a, b) by elimination of the migration term that

$$z_s = 1 + \left(\frac{y_0 - y_s}{y_0}\right) y_s \tag{4.147}$$

which can be inserted into (4.138 b) to yield the transcendental equation for y_s alone as

$$\left.\begin{aligned} &\nu y_s \exp[v(y_s)] = \left(\frac{y_0 - y_s}{y_0}\right) y_s + \nu \exp[-v(y_s)] \\ &\text{with} \\ &v(y_s) = \varrho - \alpha y_s - \beta \left(\frac{y_0 - y_s}{y_0}\right) y_s \\ &\qquad\qquad \text{where} \quad \varrho = \delta + \alpha - \beta\,. \end{aligned}\right\} \tag{4.148}$$

The equation (4.148) can be solved graphically by plotting both sides as functions of y.

To this point the discussion of (4.138) has been completely general, however, the equation will now be made specific by choosing *particular* but *reasonable trend parameters*:

$$\beta = \delta > 0 \quad \text{and} \quad y_0 = 1\,. \tag{4.149}$$

This choice implies that the fear parameter β of the prey equals its preference parameter δ for the open habitat 1 where the predators are lurking! A plausible explanation for this choice could be, for instance, that food for the prey is better in the dangerous habitat 1. Furthermore a specific saturation level $y_0 = 1$ is chosen.

For the case (4.149) the solution of (4.138) is particularly simple: It can be seen that

$$x_s = 1, y_s = 1, z_s = 1 \tag{4.150}$$

satisfies (4.138 a–c) or, alternatively, (4.138 c, 147, 148). The stability analysis also becomes reasonably transparent in this case. The secular equation (4.143) assumes the form

$$P(\lambda) = C_0\lambda^3 + C_1\lambda^2 + C_2\lambda + C_3 = 0 \tag{4.151}$$

with

$$\left.\begin{aligned} C_0 &= 1 \\ C_1 &= 1 - 2r \\ C_2 &= a(1 + s) - r \\ C_3 &= a(1 - r + s) \quad &&\text{where} \\ r &= \nu(2\alpha - 1) \quad &&\text{and} \\ s &= 2\nu\beta \,. \end{aligned}\right\} \tag{4.152}$$

The necessary and sufficient condition for the asymptotic stability of the singular point (4.150) is that all solutions to (4.151) have *negative real parts*. This condition is fulfilled if the Hurwitz criterion [4.12, 1] holds. Using the same notation as in [4.12]:

$$\left.\begin{aligned} &\text{a)}\ \frac{C_n}{C_0} > 0 \qquad \text{for} \quad n = 1, 2, 3 \\ &\text{b)}\ H_1 = C_1 > 0 \\ &\quad\ H_2 = C_1 C_2 - C_0 C_3 > 0 \\ &\quad\ H_3 = C_3 H_2 > 0 \,. \end{aligned}\right\} \tag{4.153}$$

The straightforward analysis of the Hurwitz criterion for the coefficients of (4.151) leads to the simple condition

$$\alpha < 1/2 \tag{4.154}$$

for the asymptotic stability of the singular point $P_s(1, 1, 1)$.

In the following Figs. 4.17–19 results are presented for three different values of the clustering parameter α for the constant parameters $\beta = \delta = y_0 = \nu = 1$ and $a = 2$. Figures 4.17 a, 18 a, 19 a show the graphical solution of (4.148), while Figs. 4.17 b, 18 b, 19 b exhibit the time dependence of the variables $x(\tau)$, $y(\tau)$ and $z(\tau)$ for the respective sets of trend parameters. Clearly, Fig. 4.17 b shows a damped time dependence of the variables approaching the stable focus $P_1(1, 1, 1)$ for $\alpha = 0$. Figure 4.18 b represents the case of a critical slowing-down of the oscillatory approach to the focus $P_1(1, 1, 1)$ which becomes unstable for $\alpha = 0.5$. Figure 4.19 b shows a *limit cycle behaviour*: Independent of the initial conditions all the variables $x(\tau)$, $y(\tau)$ and $z(\tau)$ approach a regular pulsatory time dependence. The singular point $P_1(1, 1, 1)$ has become unstable for $\alpha = 1.5$.

An interpretation for this phase transition in the behaviour of the prey-predator populations could be as follows:

For a small clustering trend value of α for $\mathcal{P}_\delta$ an equilibrium situation is reached in which a steady stream of prey from the refuge to the open habitat leads to a stationary number of prey in both habitats as well as of the predators living on the prey in the open habitat.

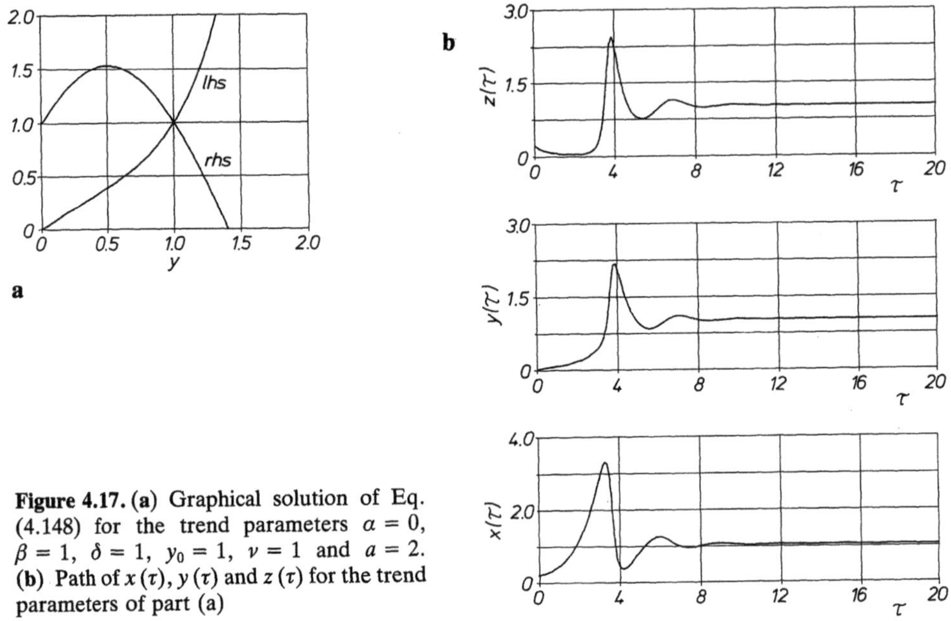

Figure 4.17. **(a)** Graphical solution of Eq. (4.148) for the trend parameters $\alpha = 0$, $\beta = 1$, $\delta = 1$, $y_0 = 1$, $\nu = 1$ and $a = 2$. **(b)** Path of $x(\tau)$, $y(\tau)$ and $z(\tau)$ for the trend parameters of part (a)

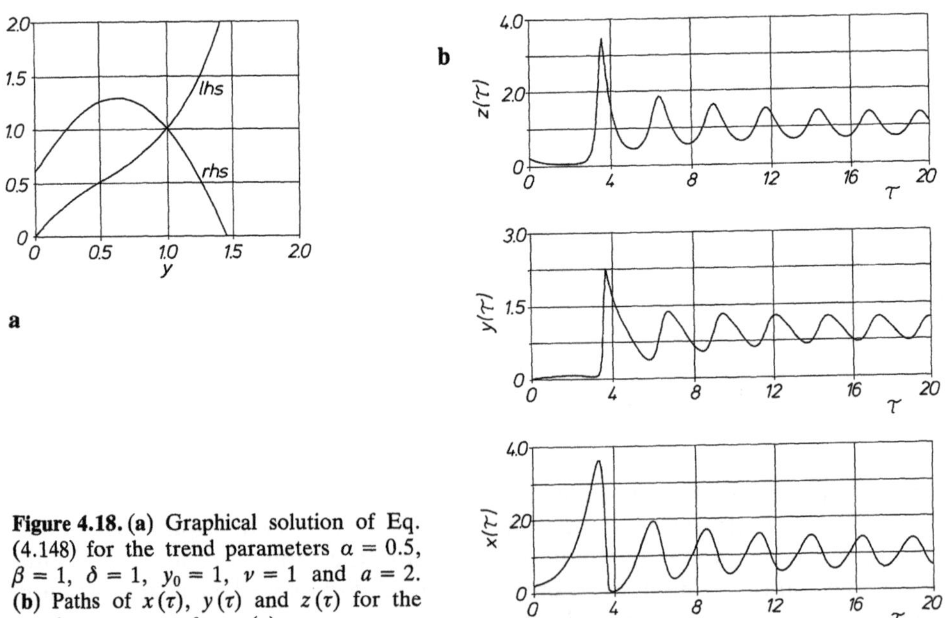

Figure 4.18. **(a)** Graphical solution of Eq. (4.148) for the trend parameters $\alpha = 0.5$, $\beta = 1$, $\delta = 1$, $y_0 = 1$, $\nu = 1$ and $a = 2$. **(b)** Paths of $x(\tau)$, $y(\tau)$ and $z(\tau)$ for the trend parameters of part (a)

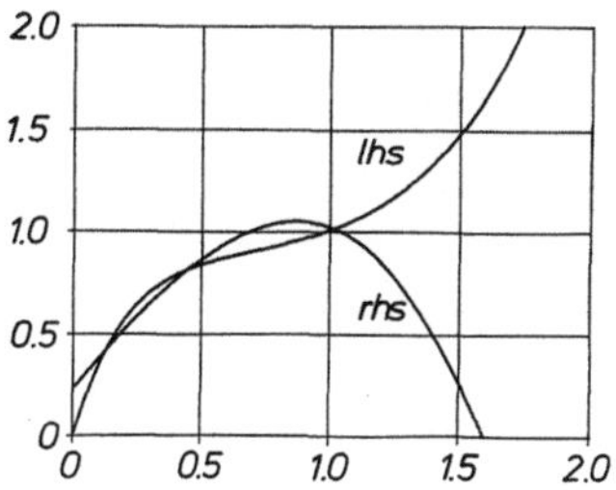

a

b

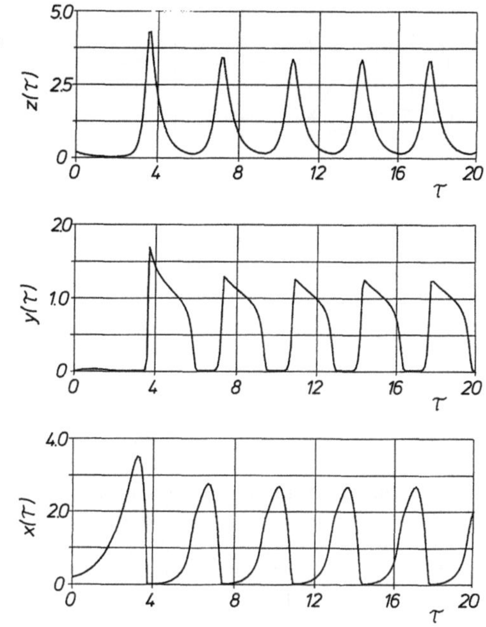

Figure 4.19. (**a**) Graphical solution of Eq. (4.148) for the trend parameters $\alpha = 1.5$, $\beta = 1$, $\delta = 1$, $y_0 = 1$, $\nu = 1$ and $a = 2$. (**b**) Paths of $x(\tau)$, $y(\tau)$ and $z(\tau)$ for the trend parameters of part (a)

For a *large clustering trend* value of α for $\mathcal{P}_\delta$, however, the migration of the prey population from the refuge habitat 2 (where it recovers but finally tends to move into the open habitat 1) takes place periodically in *swarms* and thus induces a periodic variation in the predator population $\mathcal{P}_\beta$.

Case II of (4.146) is also of interest: For $z_s = 0$ the migration terms in (4.138 a, b) can again be eliminated to yield the relation

$$x_s = -\frac{(y_0 - y_s)}{y_0}\, y_s \tag{4.155}$$

which can be inserted in (4.138 b) in order to obtain the transcendental equation for y_s alone

$$\left.\begin{aligned} &\left(\frac{y_0 - y_s}{y_0}\right)\exp[-w(y_s)] + \exp[w(y_s)] = \frac{1}{\nu}\left(\frac{y_0 - y_s}{y_0}\right) \\ &\text{with} \\ &w(y_s) = \delta - \alpha y_s\left(1 + \frac{y_0 - y_s}{y_0}\right). \end{aligned}\right\} \tag{4.156}$$

The stability analysis then shows that the one stationary point given by (4.155, 156) and $z_s = 0$ is stable. Figure 4.20 a, b show the graphical solution of (4.156) and the time dependence of $x(\tau)$, $y(\tau)$ and $z(\tau)$ for the trend parameter set $\{\delta = 0, \beta = 1, \nu = 1, y_0 = 1, \alpha = 1, a = 2\}$.

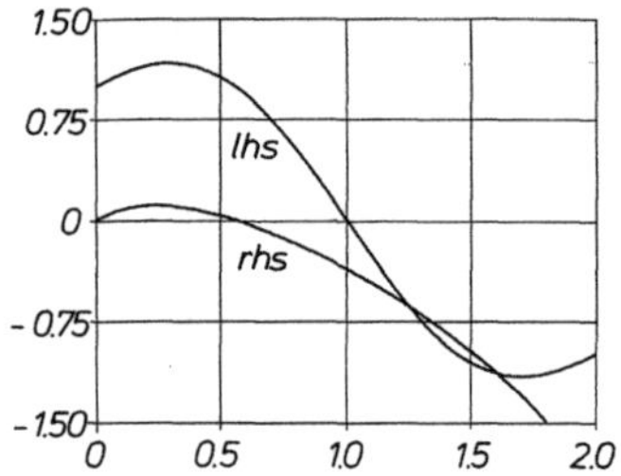

a

b

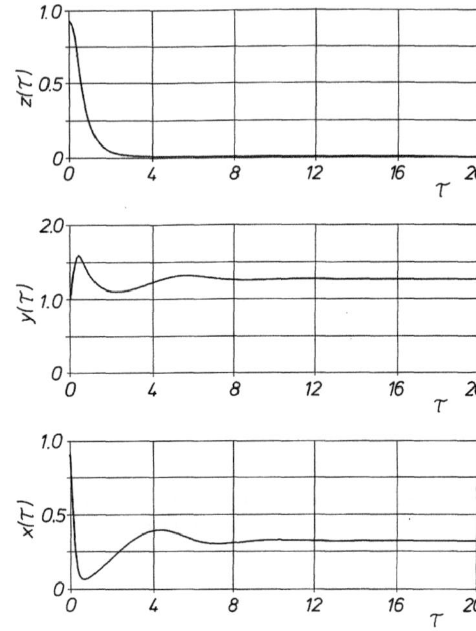

Figure 4.20. (**a**) Graphical solution of Eq. (4.156) for the trend parameters $\alpha = 1.0$, $\beta = 1$, $\delta = 0$, $y_0 = 1$, $\nu = 1$ and $a = 2$. (**b**) Paths of $x(\tau)$, $y(\tau)$ and $z(\tau)$ for the trend parameters of part (a)

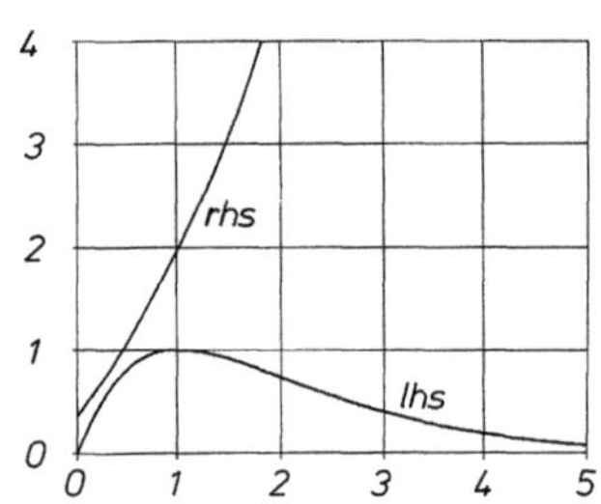

a

b

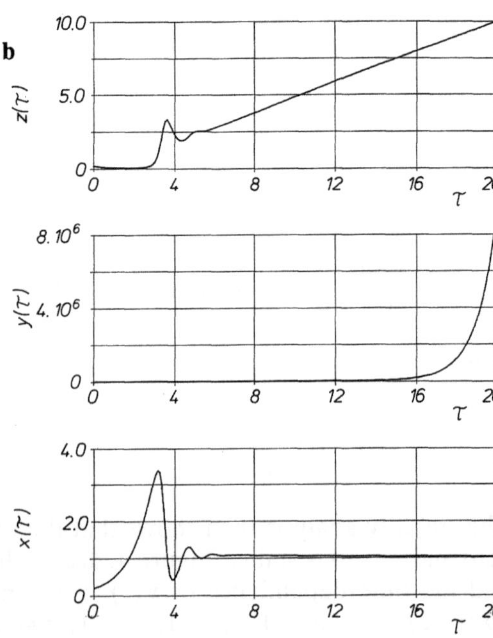

Figure 4.21. (**a**) Graphical solution of Eq. (4.158) for the trend parameters $\alpha = 0$, $\beta = 1$, $\delta = 2$, $\nu = 1$ and $a = 2$. (**b**) Paths of $x(\tau)$, $y(\tau)$ and $z(\tau)$ for the trend parameters of part (a)

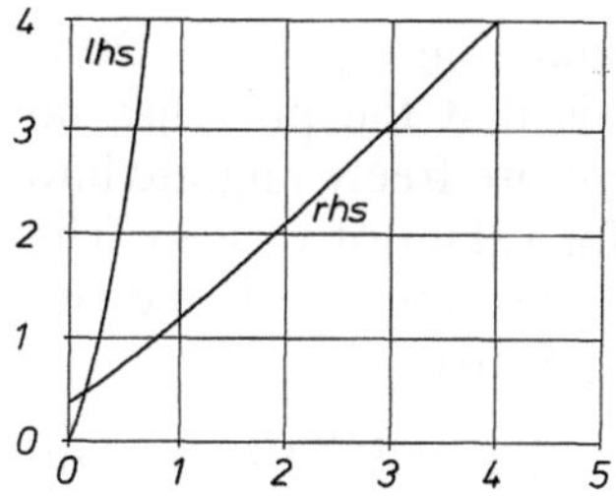

a

b

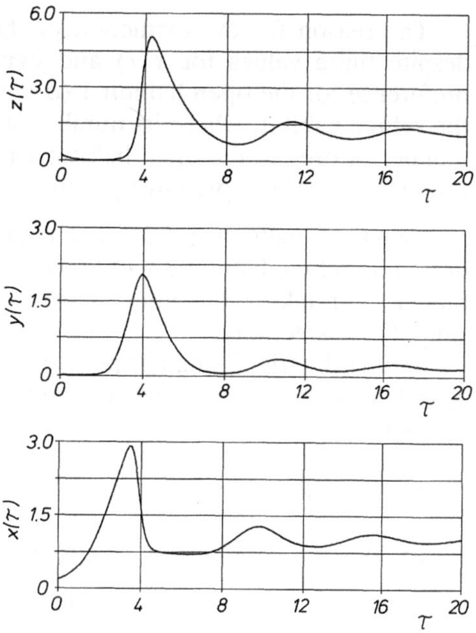

Figure 4.22. (**a**) Graphical solution of Eq. (4.158) for the trend parameters $\alpha = -2$, $\beta = 1$, $\delta = 4$, $\nu = 1$ and $a = 2$. (**b**) Paths of $x(\tau)$, $y(\tau)$ and $z(\tau)$ for the trend parameters of part (a)

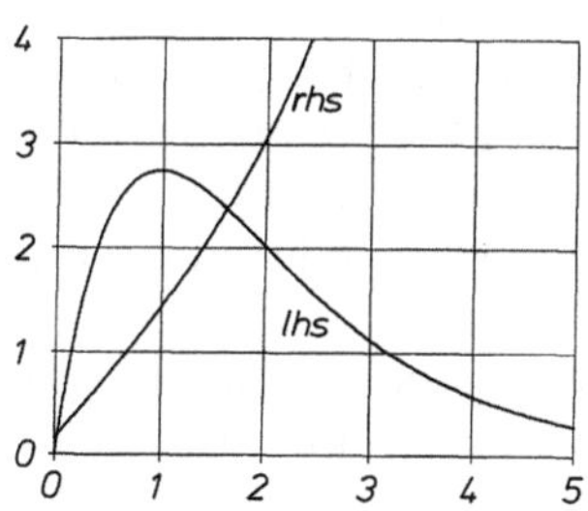

a

b

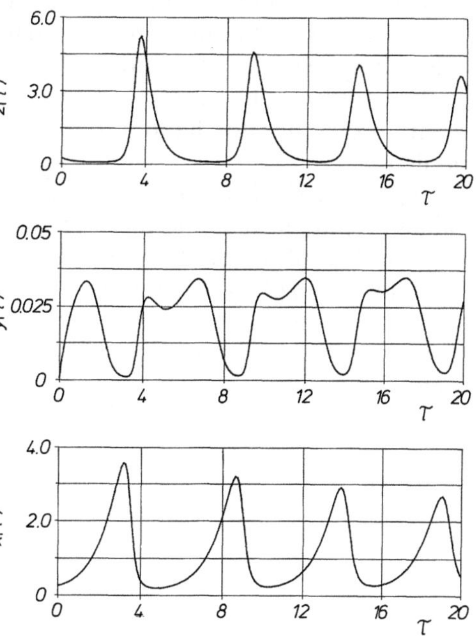

Figure 4.23. (**a**) Graphical solution of Eq. (4.158) for the trend parameters $\alpha = 1$, $\beta = 0$, $\delta = 1$, $\nu = 1$ and $a = 2$. (**b**) Paths of $x(\tau)$, $y(\tau)$ and $z(\tau)$ for the trend parameters of part (a)

The reason for the extinction of the predator population $\mathcal{P}_\beta$ (Fig. 4.20b) despite finite values for $x(\tau)$ and $y(\tau)$ of the prey is that the prey has no preference for the open habitat 1 ($\delta = 0$) and can therefore freely migrate into the refuge habitat, where its number stabilizes even for values of $y_s > y_0$. The number of prey in the open habitat 1 remains small and so the effective death rate $a(1-x)$ of the predator population always stays positive.

Case C. Finally the model with migration of the prey but *without any saturation* in its refuge habitat 2 will be considered. This case is contained in the full model after taking the limit $y_0 \to \infty$ with which the equations (4.137, 138) simplify considerably.

Setting the trivial singular point $P_0(0, 0, 0)$ aside and taking the Case I) with $x_s = 1$, with the above limit the modified equations (4.147, 148) are obtained as follows:

$$z_s = 1 + y_s \tag{4.157}$$

$$\left.\begin{aligned} &\nu y_s \exp(\varrho - \sigma y_s) = y_s + \nu \exp(-\varrho + \sigma y_s) \\ &\quad \text{with} \quad \varrho = \delta + \alpha - \beta \qquad \text{and} \quad \sigma = \alpha + \beta\,. \end{aligned}\right\} \tag{4.158}$$

The solution $z_s = 0$ which implies an extinct predator population, can therefore only occur in the trivial case $x_s = y_s = z_s = 0$. As the (graphical) solution of (4.158) can lead to 0, 1 or 2 singular points, one example of each of these cases is presented in the Figures 4.21–4.23.

In Fig. 4.21b it can be seen that both populations increase indefinitely (population explosion!) due to the fact that there is no saturation limit for the prey population in the refuge as well as a moderate preference for habitat 1.

In Fig. 4.22b the solution shows damped oscillations approaching stationary values in spite of the assumption of the unsaturated birth of the prey population in their refuge. The reason for this is an anticlustering trend together with a strong preference for habitat 1 where the prey population is reduced by the predators.

Figure 4.23b also shows damped oscillations (slowly) approaching stationary values despite the assumption of unsaturated birth. The built-in – but unjustified – lack of fear of the predator and a moderate preference of the prey population for habitat 1 explain these paths.

This discussion of a selected set of solutions of the model shows how complex even simple systems including predator-prey interaction and (non-linear) migration can be. This fact may shed a new light on the manifold of channels of survival shown by species in the course of evolution.

5. Non-Equilibrium Theory of Investment: "The Schumpeter Clock"

Economics is that social science in which the application of quantitative or semiquantitative methods has a long and well-established tradition [5.1–10]. It should therefore be a good test for the concepts developed in this book whether they can also be applied to economic problems. Interdisciplinary work of this kind, however, necessitates a formulation in the terms of the science into which the proposed model has to be embedded. Hence it was thought to be appropriate to make this chapter self-contained and as such understandable for economists without them being urged, or it being necessary for them, to read the other parts of the book. With this intention the general line of the article by *Mensch*, et al. [5.11] has been followed in formulating the essential parts of this chapter.

5.1. Introduction

5.1.1 The Relation to Preceding Concepts and Models

The interrelation of this application to economics with the general concepts formulated in preceding chapters should, however, first be made obvious to the "systematic" reader:

One of the central concepts of quantifying the "*Schumpeter* Business Cycle" [5.13] will be the investors' configuration which characterizes the group of investors (entrepreneurs, managers) with regard to their investment decisions. This investors' configuration – to be introduced in Sect. 5.2.2 – turns out to be a special case of a socio-configuration, a concept introduced in a general way in Sect. 3.1. Formally, the investors' configuration could correspond to the socio-configuration of the opinion formation model of Chap. 2, including the correspondence between their equations of motion. However, the differences between Chap. 2 and 5 are that:

1) the substantial aspect to be considered has changed from political opinions to investment behaviour in the economy,

2) the population concerned has changed from the large group of people having political opinions to the small group of entrepreneurs in a position to make investment decisions,

3) the economic decisions made are directly connected with *material variables*, namely with the "investment structure index" to be introduced in Sect. 5.2.1.

4) the trend parameters are now treated – in contrast to Chap. 2 – as *dynamic variables* obeying certain equations of motion (see, for example, the equation of motion for the investors' propensities in Sect. 5.3.2). Hence, the investors' interaction model will provide a first example for the general problem of the dynamics of trend parameters coupled to the dynamics of the socio-configuration, as already discussed in Sect. 3.3.

5.1.2 The Purpose, the Main Proposition and the Limitations of the Schumpeter Clock Model

The model presented is meant to provide a partial theory for the non-equilibrium motion of an industrial system of nations or regions. The model differs from other existing models of industrial fluctuations in many of its design principles and it is called here "the Schumpeter Clock" according to a proposal from *Goodwin* [5.12], since its moving parts, driving mechanism and control devices (variables, parameters and equations of motion), are typically Schumpeterian and not, as in the case of other models, typically neo-classical or neo-keynesian.

The Schumpeter Clock model [5.13] stresses the existent, explicitly active pushing micro-economic forces and powerful supply side checks and balances in explaining the short term non-equilibrium motions of an economy, whereas other model designs primarily take into account the less definite macro-economic forces and weak demand side checks and balances.

In particular, the model will operationalize micro-economic differences, i.e. the heterogeneity among products and production processes. These differences come into play at the sub-unit level (firms, markets, industries, sectors) of the economic system. The creation of such differences (leading to competitive advantages among rivaling producers) is the objective of the *strategic investments of entrepreneurs,* which are classified here according to their respective purposes as *"expansionary" or "rationalizing"*. Such strategic investments, notably those of the innovative type, have gained increasing influence on the growth and direction of an economy in a world of greater technological complexity, higher intensity of competition in oligopolistic markets etc., while the influence of so-called "demand disequilibrium induced investment" has – at least relatively – decreased. This tendency is systematic: Although in the future as in the past there neither can be nor was an effective supply without at least a potential demand, aspects related to the supply-side have gained more and more influence on the decision making of investors. The pressure to conform to imitative investment behaviour under some circumstances, and the subsequent need to break away from conformity by differentiation, innovation etc. under other, different circumstances, will probably increase rather than decrease in the future.

From the notion of a dynamics of the shifts between differentiation (innovation) and conformative behaviour (imitation) the main argument for the cyclicity of industrial (short-term) development presented here is derived: over the years the majority of industrial investors alternatively shifts from a predominantly expansionary investment portfolio to a predominant rationalizing investment portfolio, and vice versa, back and forth from one bias to the other, with innovators and pioneering entrepreneurs in search of monopoly profits taking the lead in the anticyclical *redirection of investment strategies*. Taken altogether, this "band-waggon" effect results in industrial fluctuations.

In the following sections equations of motion, which quantify the dynamics of economic change, will be set up and solved. Simultaneously the model design connects the relevant micro-economic and macro-economic concepts. However, the model will certainly be partial, with its focus on the theory of industrial fluctuations and in its abstraction of the impact of other substantive areas of economic theory. The following specific limitations apply:

1) *Sectoral Restrictions*. The focus is on the "Schumpeter goods sector" which is largely identical with private and public industry and those parts of agriculture and private and public services that are operating similarly to industrial organizations.

2) *Spatial Restrictions*. The focus is on major statistical units such as whole nations, states or regions.

3) *Functional Restrictions*. The focus is on the industrial investor (and innovator) and his strategic behaviour under conditions of rivalry, whereby the causes and effects of macro-economic, demand side "induced" investment are neglected. Here, the word "induced" implies either "induced by temporary macro-economic effects, such as differences between aggregate saving $S(Y_t)$ and aggregate investment $I(Y_t)$" or "induced by temporary disequilibria on the demand side, such as differences between successive levels of total consumption $C_t - C_{t-1}$ or total income $Y_t - Y_{t-1}$.

4) *Temporal Restrictions*. The focus is on the "continuity phases" in long term economic development. In the phase transition model of Kondratieff, Kuznets and Schumpeter these are the periods of recovery, boom and recession which lie inbetween major discontinuities such as depressions and clusters of basic innovations which tend to occur during depressions [5.14].

5.2 Macro- and Micro-Economic Variables of the Model and their Interdependence

5.2.1 Strategic Investment

Strategic investment plays the role of the prime mover of the economy in the model of industrial cycles presented here. How can this category of investment be defined?

In several western countries, public or private institutions collect data on the volume of Fixed Capital Investment (in plant and equipment) and Inventory (in finished or unfinished products) and on the proportions of Expansionary Investment $E(t)$, Rationalizing Investment $R(t)$ and Replacements in the Volume of Investment Expenditures $I(t)$. The usual distinction between "induced investment" and "autonomous investment", which is being made in all textbook models of the business cycle [5.2, 5, 12] cuts across the above mentioned statistical categories. For example, *Goodwin* introduced the notion of induced investment by an exclusion method: "Abstracting from trend and autonomous investment, we may state the theory as follows: today's income is greater than yesterday's income by the difference between today's investment and the saving S (Y) out of yesterday's income" [Ref. 5.5, p. 318]. Similarly, *Hicks* defined induced investment by excluding exogenous investment (triggered by technological change, change in tastes or government policy) and endogenous investment ("strategic" = growth path oriented). *Hicks* was also aware that "autonomous investment" works "through the mind of entrepreneurs" [Ref. 5.2, Foreword], which is yet another olive branch given to the neo-Austrian theory of action, but no more than that. In the model developed, "strategic" investment is largely identical with "autonomous" investment and the notion of "strategic" is considerably more extended than *Hicks*' ("growth path oriented"). At any given time an investor may have no or several strategic options. Generically though, it is assumed that the investor's strategical choice set contains only two types of alternatives, *expansionary* or *rationalizing investment projects.* According to Gold's managerial decision coefficient model there are only these two principle option types, those that are directed at either increasing sales volume S or decreasing the cost level C in the return-on-investment (r.o.i.) formula:

$$\text{r.o.i.} = \frac{\text{S} - \text{C}}{\text{output}} \cdot \frac{\text{output}}{\text{capacity}} \cdots \frac{\text{capacity}}{\text{fixed investment}} \, . \tag{5.1}$$

Thus, both expansionary (E-type) and rationalizing (R-type) investment can increase the r.o.i. by increasing the profits (S – C).

Following the above remarks the *total volume of strategic investment* [5.3]

$$I(t) = E(t) + R(t) \tag{5.2}$$

is defined as all fixed capital investment, replacement deducted, which is considered to be tactical, where

$$\left.\begin{aligned} &E(t)\text{: \textit{volume of expansionary investment} at time } t\\ &R(t)\text{: \textit{volume of rationalizing investment} at time } t\,. \end{aligned}\right\} \tag{5.3}$$

The main proposition of this chapter is that there are considerable fluctuations in the volume of expansionary and rationalizing investment as a result of entrepreneurial innovation and imitation activities. The proportional distribution of the shares of $E(t)$ and $R(t)$ in $I(t)$ shift with time. As these fluctuations take place around long term average paths $E_0(t)$ and $R_0(t)$ of the expansionary and rationalizing investment it is first appropriate to decompose $E(t)$ and $R(t)$ into

$$\left.\begin{aligned} E(t) &= E_0(t) + B(t) \\ R(t) &= R_0(t) - B(t) \end{aligned}\right\} \tag{5.4}$$

where the oscillating shift $B(t)$ around the average values E_0 and R_0 is of interest while the long term behaviour of E_0 and R_0 with time will not be considered in the following. As $E(t)$, $R(t)$ and E_0, R_0 are positive semi-definite quantities by definition, $B(t)$ can only vary within the range

$$-E_0 < B(t) < +R_0 . \tag{5.5}$$

An *investment structure index* in now defined as

$$Z(t) = \frac{E(t) - R(t)}{E(t) + R(t)} = \frac{E(t) - R(t)}{I} \tag{5.6}$$

where $Z(t)$ varies within $-1 < Z(t) < +1$. Inserting (5.4) into (5.6), $Z(t)$ decomposes into

$$Z(t) = Z_0 + z(t) = \frac{E_0 - R_0}{I} + \frac{2B(t)}{I} . \tag{5.7}$$

The performance of the Schumpeter Clock will be demonstrated by observing the non-equilibrium motion of the investment structure index $Z(t)$, or better of its fluctuating part $z(t)$, for the industrial sector in the Federal Republic of Germany over the period 1956–1979 which will be then explained by using micro-economically determined and supply side factor reinforced shifts of bias in the overall investment activities (Sect. 5.5).

5.2.2 The Investors' Configuration

In principle, an industrial firm may plan and/or undertake any number of projects at a given time t, but since the "investor" is defined as the decision making unit, one firm per project will be assumed. The total number of investment projects is large, say $2N \gg 1$, for it is convenient – but not decisive – for the equations to have an *even* number of projects. For simplicity, it will also be assumed that all projects have the same financial volume – this assumption has no effect on the mean value considerations of this chapter, a variation in the size of projects would only influence the variances of the variables in the theory.

A fictitious "neutral" investor who behaves according to the average long term investment trend will first be considered. His investment project of volume $i = I/2N$ is composed of expansionary investment e_0 and rationalizing investment r_0 as follows:

$$i = \frac{I}{2N} = e_0 + r_0 \quad \text{where} \quad e_0 = \frac{E_0}{2N} \quad \text{and} \quad r_0 = \frac{R_0}{2N} . \tag{5.8}$$

Real investors, however, behave differently to the neutral investor: There are *E-type investors* who favour the expansionary investment type against the average trend. Their projects of financial volume i are constructed as follows:

$$\left.\begin{aligned} & i = e_E + r_E \quad \text{where} \\ & \qquad e_E = e_0 + b \quad \text{and} \quad r_E = r_0 - b \quad \text{with} \quad b > 0\,. \end{aligned}\right\} \tag{5.9}$$

Evidently b is the surplus of expansionary investment in comparison with the neutral case.

On the other hand, *R-type investors* favour the rationalizing investment type in comparison with the average trend. The *E/R* type shares of their projects are such that

$$\left.\begin{aligned} & i = e_R + r_R \quad \text{where} \\ & \qquad e_R = e_0 - b \quad \text{and} \quad r_R = r_0 + b \quad \text{again with} \quad b > 0\,. \end{aligned}\right\} \tag{5.10}$$

b is now the surplus in rationalizing investment with respect to the neutral case. For simplicity, the same value for b in (5.9,10) will be taken.

At a given point in time $n_E(t)$ E-type investors and $n_R(t)$ R-type investors, whose total number is $2N$, where:

$$n_E(t) + n_R(t) = 2N \tag{5.11}$$

are assumed.

The two numbers in the vector $\{E(t),\ R(t)\}$ characterize the investment structure at time t according to investment volume and the two numbers $\{n_E(t), n_R(t)\}$ characterize the investors' strategic investment activities at time t by head count. $\{n_E(t), n_R(t)\}$ is denoted as the *investors' configuration*.

The integer

$$n(t) = \frac{n_E(t) - n_R(t)}{2} \quad \text{where} \quad -N \leq n(t) \leq N \tag{5.12}$$

increases or decreases by one if the investors' configuration changes according to transitions

$$\{n_E, n_R\} \to \{n_E + 1, n_R - 1\} \tag{5.13 a}$$

or as

$$\{n_E, n_R\} \to \{n_E - 1, n_R + 1\}\ , \tag{5.13 b}$$

i.e. if a R-type investor becomes an E-type investor or vice versa. Multiple-unit motions (band-waggon effects) within the investors' configuration are also possible, for instance,

$$\left.\begin{aligned} &\{n_E, n_R\} \to \{n_E + d, n_R - d\} \\ &\quad n \to n + d\,. \end{aligned}\right\} \tag{5.14a}$$

Such transitions may indicate series of synchronized *product* variations or innovations, as often observed and described by that heuristic regularity usually called the product life cycle. Similarly,

$$\left.\begin{aligned} &\{n_E, n_R\} \to \{n_E - d, n_R + d\} \\ &\quad n \to n - d \end{aligned}\right\} \tag{5.14b}$$

indicates series of synchronized *process* improvements or process rationalizations.

Instead of $n(t)$, a normalized variable, the *investors' configuration index,* can be employed

$$x(t) = \frac{n_E(t) - n_R(t)}{n_E(t) + n_R(t)} = \frac{n(t)}{N} \tag{5.15}$$

with $-1 \le x(t) \le 1$.

5.2.3 Strategic Investment and the Investors' Configuration

The relation between the investors' configuration index (5.15) and the investment structure index (5.6) characterizing strategic investment now follows unambiguously by combining the postulated equations. The total expansionary investment $E(t)$ and the total rationalizing investment $R(t)$ at time t are given by

$$E(t) = n_E(t)\,e_E + n_R(t)\,e_R \tag{5.16}$$

$$R(t) = n_E(t)\,r_E + n_R(t)\,r_R\,. \tag{5.17}$$

Inserting (5.8–10) as well as (5.11, 12 and 15) it follows that

$$\begin{aligned} E(t) &= [n_E(t) + n_R(t)]\,e_0 + [n_E(t) - n_R(t)]\,b \\ &= 2N e_0 + 2n(t)\,b = E_0 + 2Nbx(t) \end{aligned} \tag{5.18}$$

and

$$\begin{aligned} R(t) &= [n_E(t) + n_R(t)]\,e_0 - [n_E(t) - n_R(t)]\,b \\ &= 2Nr_0 - 2n(t)\,b = R_0 - 2Nbx(t)\,. \end{aligned} \tag{5.19}$$

Using (5.18, 19) in the definition (5.6) or (5.7) of $Z(t)$ results in

$$Z(t) = \frac{E_0 - R_0}{I} + \frac{4Nb}{I} x(t) = Z_0 + z(t) \qquad \text{or} \tag{5.20}$$

$$z(t) = \frac{4Nb}{I} x(t) = gx(t) \qquad \text{with} \quad g = \frac{4Nb}{I}. \tag{5.21}$$

The result (5.21) shows that the fluctuating part $z(t)$ of the investment structure index $Z(t)$ is proportional to the investors' configuration index $x(t)$. Thus oscillations of the investors' configuration will show up in oscillations of the investment structure index.

The discussion of the investor's strategic choice set indicated that the model is consistent with the traditional profit seeking hypothesis. The investigation of the investors' configuration now reveals that the model is also consistent with the rational expectation hypothesis: "Rational" expectations are model based, and the life cycle model and progress function model are frequently used by investors to form expectations about the timing [5.15, 16] of their rivals' next improvements, i.e. of the current best practice frontier. *Rosenberg* [5.17] in his article on *Technological Expectations* also pointed out the rationality – under conditions of rivalry – of speeding-up or delaying product or process innovations in the expectation, for example, of breakthroughs and rapid successions of product or process innovations, which would hit a market as a series of capacity expanding or cost reducing investments. For example, anticipating an imitation of cost reducing investment is rational under conditions of rivalry, and as in the "awareness context" [5.18] of industrial investors "a firm must incur some positive expense just to maintain a constant level of production cost or efficiency" [Ref. 5.19, p. 1187]; this is especially true in times of cost inflation when the goal is one of "reducing cost in total, not particular costs such as labour costs or capital costs" [Ref. 5.20, p. 44]. This micro-economic theory of macro-economic contraction also suggests that – in times of cost inflation in which industrial firms try to combat a strong rationalization bias in their overall investment activities – heavy investment does not create higher employment, as the Phillips curve would suggest, but "stagflation".

Turning now to the issue of the generation of macro-economic cycles from what seems to be a condition of micro-economic chaos it can be shown that the single unit motions of the investors' configuration form a collective scheduled and synchronized pattern under appropriate assumptions about the interaction between investors.

5.3 Design of the Investors' Interaction Model

The short term industrial investment cycles, it will be assumed, come about through dynamic interaction between industrial investors, or between two interrelated components of investment behaviour: a) the *decision behaviour* of managers and entrepreneurs who are making strategic choices about the kind of industrial investments to be implemented, and b) the actual total volume and composition of *realized investments*. As these components are embodied in the investors' configuration, changes in the numbers $n_E(t)$ and $n_R(t)$ of investors undertaking expansionary or rationalizing investments can be taken as a proxy for the real socio-economic causes of the time path of all industrial activity. Changes in the investors' configuration $\{n_E(t), n_R(t)\}$ result in changes in the rate and direction of industrial investment as indicated by the investment structure index $Z(t)$; and these changes come about as investors change their propensities for E- and R-type investments in view of market opportunities and the revealed preferences of other investors as well as other supply side conditions.

Thus, changes in the industrial economy can be formulated as equations of motion of these two components of investment behaviour. The *equation of motion for the investors' configuration* and the *equation of motion for the investors' propensities,* as presented in the next two sections, constitute essential parts in the design of the Schumpeter Clock model being presented.

5.3.1 The Equation of Motion for the Investors' Configuration

It has been seen that the transition from one investors' configuration $\{n_E, n_R\}$ to another can be a single unit motion connected with a product innovation (5.13 a) or a process innovation (5.13 b) of one investor, or a multiple unit motion (5.14 a and b) which is most often an imitation process. As innovation always and imitation sometimes is investment under uncertainty, it can be suspected that a deterministic modelling of the motion of the investors' configuration will not succeed. But since all investment is risky, the stochastic approach *is* appropriate. Therefore the well founded master equation formulation will be adopted.

a) The Transition Probabilities. The micro-economic approach to changes in the investors' configuration $\{n_E, n_R\}$ adopted incorporates the notion of individual transition probabilities (per unit of time) for investors turning from an R-type investment to an E-type investment, and vice versa. These individual transition probabilities are denoted by:

$$p_{E \leftarrow R}[n_E, n_R] \equiv p_\uparrow(n) = \text{probability per unit time for turning from } R\text{-type to } E\text{-type investment} \tag{5.22}$$

$$p_{R \leftarrow E}[n_E, n_R] \equiv p_\downarrow(n) = \text{probability per unit time for turning from } E\text{-type to } R\text{-type investment.} \tag{5.23}$$

The individual transition probabilities yield the total probabilities for changes of the entire investors' configuration. The transition

$$\{n_E, n_R\} \rightarrow \{n_E + 1, n_R - 1\}$$

takes place with the total transition probability

$$w_{\uparrow}(n) = n_R p_{\uparrow}(n) = (N-n) p_{\uparrow}(n) \tag{5.24}$$

as there are n_R investors who could turn from R-type to E-type investment with individual probability $p_{\uparrow}(n)$. Analogously, the transition

$$\{n_E, n_R\} \rightarrow \{n_E - 1, n_R + 1\}$$

takes place with the total transition probability

$$w_{\downarrow}(n) = n_E p_{\downarrow}(n) = (N+n) p_{\downarrow}(n) \tag{5.25}$$

as there are n_E investors who could turn from E-type to R-type investment with individual probability $p_{\downarrow}(n)$. These transition probablities (5.24, 25) of single unit motions of the investors' configuration $\{n_E, n_R\}$ are the prerequisites for formulating the master equation.

b) The Master Equation for the Motion of $p(n; t)$. The master equation does not directly describe motions of the investors' configuration, but the motion of the *probability distribution* over the investors' configuration. This distribution function is denoted as

$$p[n_E, n_R; t] \equiv p(n; t) \qquad \text{for} \quad -N \leq n \leq N . \tag{5.26}$$

By definition the value of this function of n and t is the probability that at time t the investors' configuration $\{n_E, n_R\}$ is realized; the configuration relevant variable of the function $p(n; t)$ is $n = (n_E - n_R)/2$. Because one of the configurations $\{n\}$ is *always* realized, the sum of the probabilities has to be 1:

$$\sum_{n=-N}^{N} p(n; t) = 1 . \tag{5.27}$$

The master equation is the equation of motion for $p(n; t)$. It is derived by a simple probability balance consideration: the probability $p(n; t)$ of the configuration n can *increase* by means of transitions from either one of the neighbouring configurations $(n-1)$ or $(n+1)$ into configuration n, while $p(n; t)$ may simultaneously *decrease* by transitions from configuration n into these neighbouring configurations. The transition probabilities per time unit for these processes have alrady been introduced in (5.24, 25). Including them into the balance consideration immediately leads to the master equation

$$\begin{aligned}\frac{dp(n; t)}{dt} = {} & [w_{\uparrow}(n-1) p(n-1; t) + w_{\downarrow}(n+1) p(n+1; t)] \\ & - [w_{\uparrow}(n) p(n; t) + w_{\downarrow}(n) p(n; t)] .\end{aligned} \tag{5.28}$$

The first term on the rhs of (5.28) describes the probability flux per time unit *into* the configuration n and the second term the flux *from* the configuration n.

The master equation (5.28) represents $2N + 1$ coupled linear difference differential equations for $p(n; t)$ with $n = -N, -N + 1, \ldots, N - 1, N$, which are not easily solved in the general case when N is large. However, if it is assumed that the $p(n; t)$ are sharply peaked and uni-modal around their mean values $\langle n \rangle_t$, then a set of closed equations of motion for all the moments $\langle n^k \rangle_t$, $(k = 1, 2, 3 \ldots)$ of the distribution function can be derived. The moments are defined by

$$\langle n^k \rangle_t = \sum_{n=-N}^{N} n^k p(n; t) . \tag{5.29}$$

The analysis of the equations of motion for the higher moments $(k = 2, 3, \ldots)$ plays a considerable role if the model is used for predictive purposes. Here the consideration is restricted to descriptive modelling and to the equation of motion for the mean value $\langle n \rangle_t$ only, i.e. to the case $k = 1$ in (5.29)

$$\langle n \rangle_t = \sum_{n=-N}^{N} n p(n; t) . \tag{5.30}$$

c) The Mean Value Equation for the Motion of n. Taking the time derivative of (5.30) and inserting the master equation (5.28) in the rhs, some simple rearrangements of the sums leads to the exact relation:

$$\frac{d\langle n \rangle_t}{dt} = \sum_{n=-N}^{N} n \frac{dp(n; t)}{dt} = \sum_{n=-N}^{N} [w_\uparrow(n) - w_\downarrow(n)] p(n; t) . \tag{5.31}$$

The rhs of (5.31) is the mean value $\langle w_\uparrow(n) - w_\downarrow(n) \rangle_t$ of $w_\uparrow(n) - w_\downarrow(n)$. Equation (5.31) is not a closed equation for the mean value $\langle n \rangle_t$ because the full distribution $p(n; t)$ has to be known in order to calculate $d\langle n \rangle_t/dt$. But under the above assumption that $p(n; t)$ remains sharply peaked and uni-modal around its mean value, an approximate closed equation of motion for the mean value $\langle n \rangle_t$ can be obtained as

$$\frac{d\langle n \rangle_t}{dt} = w_\uparrow(\langle n \rangle_t) - w_\downarrow(\langle n \rangle_t) . \tag{5.32}$$

d) The Mean Value Equation for the Motion of $x(t)$. The mean-value equation for x, the investors' configuration index, upon going over to the variable $x = n/N$, see (5.15), in (5.32), is obtained as:

$$\frac{d\langle x \rangle_t}{dt} = K(\langle x \rangle_t) \tag{5.33}$$

where *the driving force* $K(\langle x \rangle_t)$ of $\langle x \rangle_t$ is

$$K(\langle x\rangle) = \frac{1}{N}[w_\uparrow(\langle n\rangle) - w_\downarrow(\langle n\rangle)]$$

$$= (1-\langle x\rangle)p_\uparrow(N\langle x\rangle) - (1+\langle x\rangle)p_\downarrow(N\langle x\rangle)\,. \tag{5.34}$$

From this point on the clumsy mean value brackets will be omitted and $x(t)$ written for $\langle x\rangle_t$, thus (5.33) is

$$\frac{dx(t)}{dt} = K(x(t))\,. \tag{5.35}$$

In the Schumpeter Clock design the driving force is K which flows from the current investors' configuration $\{n_E(t), n_R(t)\}$ as represented by the state variable $x(t)$ and is mediated by force shaping parameters δ and κ which will now be specified.

e) The Driving Force $K(x)$ of x. The driving force of x in the interaction model is K. According to (5.34), K depends on the actual configuration $x(t)$ of all the actors (investors) in the system at time t and the transition probabilities for change of this configuration, $p_\uparrow(x)$ and $p_\downarrow(x)$, which in turn depend upon all investors' propensities, and which will be parametrized by δ and κ as follows:

δ: *"the alternator"* represents the investor's switch in his preference for either E- or R-type investments under the given circumstances. In the specification of transition probabilities the alternator δ is chosen as a parameter obeying the following rule: A positive δ implies that transitions to E-type investment are favoured; a negative δ favours transitions to R-type investment. The alternator $\delta(t)$ is time dependent and obeys an equation of motion (see Sect. 5.3.2) whose role in the generation of cyclic (Schumpeter Clock) motion will been seen later.

κ: *"the coordinator"* represents the individual investor's interaction intensity under the given circumstances; in other words κ represents the investor's inclination to conform to other investors' behaviour. The coordination effect will become visible as a synchronization and imitation of investments undertaken by various investors.

Thus the driving force K, see (5.34), may be written as a function of x, δ and κ as

$$K(x;\delta,\kappa) = (1-x)p_\uparrow(x;\delta,\kappa) - (1+x)p_\downarrow(x;\delta,\kappa)\,. \tag{5.36}$$

A constructive specification of these transition probabilities which is fairly general and which has already been shown to be sufficiently flexible in the opinion formation model of Chap. 2, reads

$$\left.\begin{aligned} p_\uparrow(x;\delta,\kappa) &= \nu\exp(\delta+\kappa x)\\ p_\downarrow(x;\delta,\kappa) &= \nu\exp[-(\delta+\kappa x)]\,,\end{aligned}\right\} \tag{5.37}$$

where ν is a scalar used in actual computations for time scaling purposes. It can be easily seen from (5.37) that the parameters δ and κ fulfill their above specified role as "alternator" and "coordinator".

Inserting (5.37) in (5.36) and using the standard definitions of the hyperbolic sine and cosine:

$$\left.\begin{aligned} \sinh(y) &= 1/2\,[\exp(y) - \exp(-y)] \\ \cosh(y) &= 1/2\,[\exp(y) + \exp(-y)] \end{aligned}\right\} \tag{5.38}$$

a specific expression for the driving force

$$K(x;\,\delta,\kappa) = 2\nu\,[\sinh(\delta + \kappa x) - x\cosh(\delta + \kappa x)] \tag{5.39}$$

is obtained which in turn yields an explicit form for the mean value equation of motion of the investors' configuration index

$$\frac{dx(t)}{dt} = K(x;\,\delta,\kappa) \tag{5.40}$$

by using (5.39) on the rhs of (5.40).

f) A Discussion of the Dynamic Behaviour of $x(t)$. The dynamic behaviour of x can also be described in terms of a "potential" which stands in close relation to the driving force K of x

$$K(x;\,\delta,\kappa) = -\frac{\partial V(x;\,\delta,\kappa)}{\partial x}, \tag{5.41}$$

which may be regarded as being the pace maker of the Schumpeter Clock. Integrating the force (5.39) yields the *"configuration development potential"*

$$V(x;\,\delta,\kappa) = \frac{2\nu}{\kappa^2}\,[\kappa x\sinh(\delta + \kappa x) - (1+\kappa)\cosh(\delta + \kappa x)] \tag{5.42}$$

In the upper part of Fig. 5.1 numerical values for $V(x;\,\delta,\kappa)$ are depicted:

(a) for $\delta = 0$ and different values of κ
(b) for $\kappa = 1{,}5$ and different of values δ,

whereas in the lower part of Fig. 5.1, c, d, the quantified driving forces $K(x;\,\delta,\kappa)$ associated with the cases a and b, respectively, are shown.
Comparing Fig. 5.1 a with c and b with d for the same sets of parameters κ, δ it is seen that the force $K(x;\,\delta,\kappa)$ drives x into the minima of the potential $V(x;\,\delta,\kappa)$. In the Case a with $\kappa = 1.5$, for instance, the motion ends at the positive minimum $x = x_+$ (expansionary bias of the investors' configuration index x) or at the negative minimum at $x = x_-$ (rationalization bias of x). In the Case b with $\delta = 0.3 > \delta_c$ the potential has one minimum only at $x = x_+$ and the motion ends with the strong expansionary bias $x = x_+$.

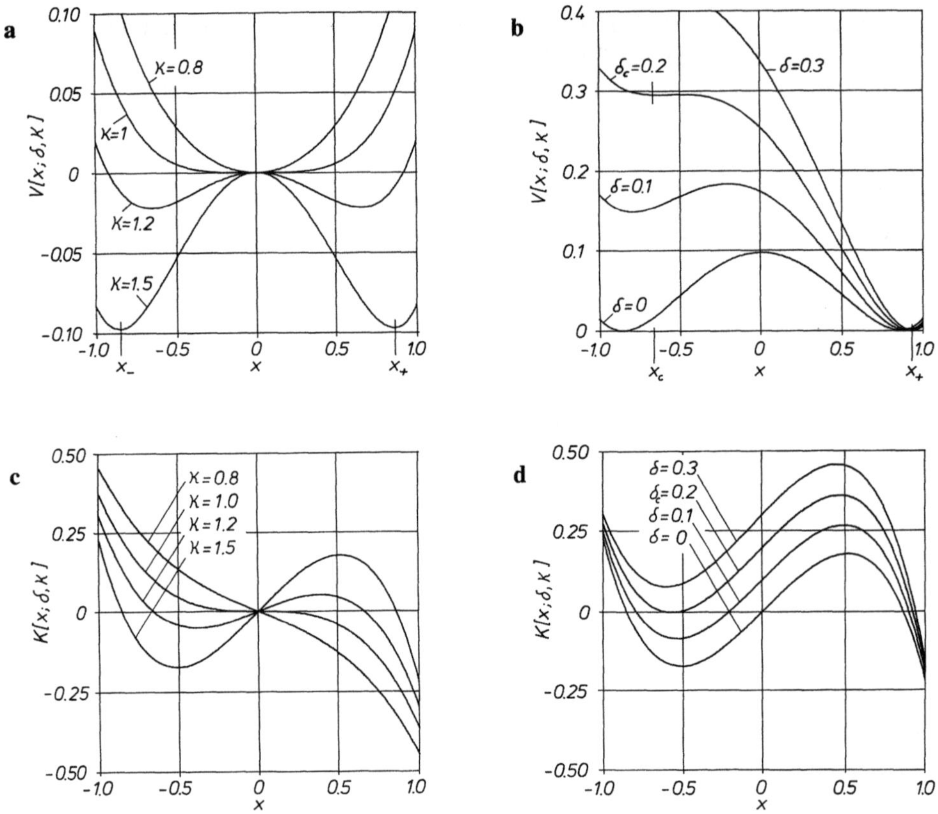

Fig. 5.1. (**a**, **b**) Typical graphs of the configuration development potential $V(x;\ \delta,\ \kappa)$. (**c**, **d**) The associated driving forces for $x(t)$:

$$K(x;\ \delta,\ \kappa) = -\frac{\partial V(x;\ \delta,\ \kappa)}{\partial x}$$

So far it might be guessed that the motion of the investors' configuration index x always ends at a stable (expansionary or rationalizing) bias and that no Schumpeter Clock begins to tick. But this is only true as long as δ and κ are kept *constant* with time. In Sect. 5.3.2, however, an equation of motion for the alternator $\delta(t)$ will be set up. Becoming then a *dynamic variable*, the alternator $\delta(t)$ shifts and transforms the whole shape of the configuration development potential $V(x;\ \delta, \kappa)$ and of the driving force $K(x;\ \delta, \kappa)$. Hence x may never come to rest; in fact a periodicity in the motion of $x(t)$ can result from the coupling of the motion of $\delta(t)$ with that of $x(t)$ and so major shifts in the variable x in the course of this coupled motion have to be expected. In Fig. 5.1 b it can be seen that for given κ the shape of the potential varies from a two minima to a one minimum form as δ grows from 0 to $\delta > \delta_c$. The transition takes place at a critical value $\delta = \delta_c$ given by

$$\cosh^2[\delta_c - \sqrt{\kappa(\kappa-1)}] = \kappa \tag{5.43}$$

for which the left hand minimum of the potential disappears at

$$x_c = -\sqrt{\kappa(\kappa-1)} \tag{5.44}$$

Because of the symmetry of the potential

$$V(-|x|, -|\delta|, \kappa) = V(|x|, |\delta|, \kappa) \tag{5.45}$$

a similar transition takes place for $\delta = -|\delta|$ for which the right hand minimum disappears at $x = |x_c|$. If $x(t)$ is considered starting at the left minimum of $V(x; \delta < \delta_c, \kappa)$ – i.e. with a rationalizing bias – and if $\delta(t)$ moves from $\delta < \delta_c$ to values $\delta > \delta_c$ so that the left minimum of V disappears, $x(t)$ will quickly swing to the remaining right-hand minimum of $V(x; \delta > \delta_c, \kappa)$ at $x > 0$ (expansionary bias). This in turn can induce the motion of the alternator $\delta(t)$ to negative values $\delta(t) < -|\delta_c|$ leading to a sudden downswing of $x(t)$ towards a rationalizing bias, and so on. It can be seen that the coupled dynamics of $x(t)$ and $\delta(t)$ in principle can explain the upswings and downswings of $x(t)$. In the next section the function of a time dependent alternator will be explained in economic terms and its equation of motion will be established.

5.3.2 The Equation of Motion for the Investors' Propensities

a) The Function of the Alternator. The alternator δ, which is the investors' strategic choice parameter in the force $K(x; \delta, \kappa)$ and which is the variable for which an equation of motion is going to be introduced, plays the role of a "trend setting function":

If the majority of investors tends to maximize profits at a given point in time by expanding (rationalizing) their business operations so that $x(t) > 0$ ($x(t) < 0$) then some innovators or pioneers (trend setters) will try to improve their market position by adopting a non-conformist strategy in an attempt to capture quasi-rents due to differentiation. When an upswing is well under way due to expansionary investments undertaken by a majority of investors, these trend setters tend to redirect their effort and to start pushing back the cost frontier by means of cost reducing investments, thereby forcing others to imitate and also undertake rationalizing investments in the expectation of further cost reductions along the progress function descriptive of the least cost combination in the branch of industry under observation.

And at other times when a downswing is well under way due to the contractionary effects of rationalizing investments undertaken by a majority of investors, the trend setters start moving towards the quality section of the best practice frontier, introducing better products and implementing investment plans for expanding facilities and thereby forcing others again to imitate, in this case their expansionary and quality updating behaviour, thus creating the synchronization to be observed in the occurrance of product life cycles. Since

Adam Smith economists have viewed these entrepreneurial actions as one of the main sources of the wealth of nations, namely, the occurence of social and economic benefits due to differentiation.

b) The Equation of Motion for the Alternator $\delta(t)$. The equation of motion for the alternator $\delta(t)$, which describes in aggregate terms the differentiation activities of entrepreneurs in various fields of industrial investment, should generate switches under the circumstances stipulated in a) above. A suitable specification of this dynamic behaviour is

$$\frac{d\delta(t)}{dt} = \mu[\delta_0 - \delta(t)]\exp[-\beta x(t)] - \mu[\delta_0 + \delta(t)]\exp[\beta x(t)]$$

$$\text{with} \quad \mu > 0,\ \beta > 0 \quad \text{and} \quad \delta_0 > 0. \tag{5.46}$$

The mathematical implication of (5.46) is such that: For $x = 0$, $\delta(t)$ relaxes towards $\delta = 0$; but for $x(t) > 0$ the rhs of (5.46) represents a strong restoring force so that $\delta(t)$ moves towards $-\delta_0$. The negative δ in turn leads to a force $K(x; \delta, \kappa)$, see (5.39), driving x to negative values. Vice versa, starting from $x(t) < 0$, the rhs of (5.46) yields a restoring force for a change in $\delta(t)$ towards $+\delta_0$; the positive δ, in turn, produces a force $K(x; \delta, \kappa)$ driving x to positive values again. It is seen that (5.46) correctly describes the alternating dynamics of the strategic choice parameter $\delta(t)$ in terms of a non-conformist reaction to the investors' configuration index $x(t)$.

In order to achieve the alternating effect the parameter β, the *trend reversal speed parameter,* has to be much larger than unity. The *strategic flexibility parameter* μ describes the flexibility of the investors in turning their strategies from expansionary to rationalizing, and vice versa, whereas *the strategic choice amplitude* δ_0 is an operative scaling constant.

Using the definitions (5.38) of hyperbolic functions the general form of the equation of motion for δ is obtained from (5.46) as

$$\frac{d\delta(t)}{dt} = -2\mu\delta_0 \sinh[\beta x(t)] - 2\mu\delta(t)\cosh[\beta x(t)]\,. \tag{5.47}$$

Introducing δ_1, a *strategy bias parameter*, which is positive or negative if the entire trend period is heavily biased towards expansion (as the 1950's were) or towards rationalization (as the 1970's were), (5.47) can be modified and generalized to a complete equation of motion for the alternator which is both stable enough to hold the ongoing trend over considerable periods of inertia and flexible enough to change expeditiously with the investors' propensities:

$$\frac{d\delta(t)}{dt} = -2\mu\{\delta_0 \sinh[\beta x(t)] + [\delta(t) - \delta_1]\cosh[\beta x(t)]\}\,. \tag{5.48}$$

Analogous to $K(x; \delta, \kappa)$, $L(x; \delta, \kappa)$ is denoted as the *strategy reformulation driving force* by writing the equation of motion for the alternator in the form:

$$\frac{d\delta(t)}{dt} = L(x; \delta, \kappa) \tag{5.49}$$

with

$$L(x; \delta, \kappa) = -2\mu\,[\delta_0 \sinh(\beta x) + (\delta - \delta_1)\cosh(\beta x)] \tag{5.50}$$

and denoting the parameters as follows

β: *trend reversal speed parameter,*
μ: *strategic flexibility parameter,*
δ_0: *strategic choice amplitude,*
δ_1: *strategy bias parameter.*

5.3.3 The Closed Set of Equations of Motion

Collecting the equations of motion for the investors' configuration index $x(t)$ (5.40) and the equation of motion for the alternator $\delta(t)$ (5.49) and introducing scaled time τ as

$$\tau = 2\nu t \tag{5.51}$$

and γ, a scaled strategic flexibility parameter, as

$$\gamma = \mu/\nu \tag{5.52}$$

scaled forms of the equations of motion

$$\frac{dx(\tau)}{d\tau} = \hat{K}(x; \delta, \kappa) \tag{5.53}$$

$$\frac{d\delta(\tau)}{d\tau} = \hat{L}(x; \delta, \beta) \tag{5.54}$$

are obtained with the (scaled) driving force $\hat{K}$ as

$$\begin{aligned}\hat{K}(x; \delta, \kappa) &= \sinh(\delta + \kappa x) - x\cosh(\delta + \kappa x)\\ &\equiv [\tanh(\delta + \kappa x) - x]\cosh(\delta + \kappa x)\end{aligned} \tag{5.55}$$

and the (scaled) strategy reformulation driving force $\hat{L}$ as

$$\begin{aligned}\hat{L}(x; \delta, \beta) &= -\gamma\,[\delta_0 \sinh(\beta x) + (\delta - \delta_1)\cosh(\beta x)]\\ &\equiv -\gamma\,[\delta_0 \tanh(\beta x) + (\delta - \delta_1)]\cosh(\beta x)\,.\end{aligned} \tag{5.56}$$

It is worthwhile to note that the central equations of the model (5.53, 54) are invariant under the transformations

$$x \to -x, \delta \to -\delta \text{ and } \delta_1 \to -\delta_1 \,. \tag{5.57}$$

This invariance condition also shows up in the symmetry condition (5.45) of the potential $V(x;\, \delta, \kappa)$.

5.4 Structural Analysis of the System of Equations

The analysis starts with the two coupled autonomous non-linear differential equations (5.53, 54) set up in Sect. 5.3.3 as a result of the design of the investors' interaction model of Sect. 5.3.

5.4.1 The Singular Points of the Equations of Motion

The first part of the analysis consists in asking how many singular points $\bar{P}(\bar{\delta}, \bar{x})$ of the equations of motion exist? *Singular points* are defined as points in the δ-x plane where the motion comes to rest, i.e. where

$$\left.\frac{dx}{d\tau}\right|_{\bar{P}} = \hat{K}(\bar{x}, \bar{\delta}) = 0 \tag{5.58}$$

and

$$\left.\frac{d\delta}{d\tau}\right|_{\bar{P}} = \hat{L}(\bar{x}, \bar{\delta}) = 0 \,. \tag{5.59}$$

Inserting (5.55 and 56) into (5.58 and 59), respectively, it is seen that the coordinates of a singular point $\bar{P}(\bar{\delta}, \bar{x})$ simultaneously have to satisfy the two transcendental equations

$$F_1(\bar{\delta}, \bar{x}) \equiv [\tanh(\bar{\delta} + \kappa\bar{x}) - \bar{x}] = 0 \tag{5.60}$$

$$F_2(\bar{\delta}, \bar{x}) \equiv [\delta_0 \tanh(\beta\bar{x}) + (\bar{\delta} - \delta_1)] = 0 \,. \tag{5.61}$$

These equations are easily solved graphically: the functions $F_1(\delta, x) = 0$ and $F_2(\delta, x) = 0$ can be represented in the δ-x plane and their intersection points, which are the singular points of the differential equations (5.33, 54), determined. Furthermore, it follows from the definitions that the graph of (5.60) is the locus of all horizontal, the graph of (5.61) the locus of all vertical fluxlines of the equations (5.53, 54). Restricting the analysis for the moment to the slightly simplified case of vanishing strategic choice bias, i.e. for $\delta_1 = 0$, Fig. 5.2 shows a typical graph of (5.61) and three typical graphs a, b and c of (5.60). As seen from Fig. 5.2 there can exist one, three or even five singular points.

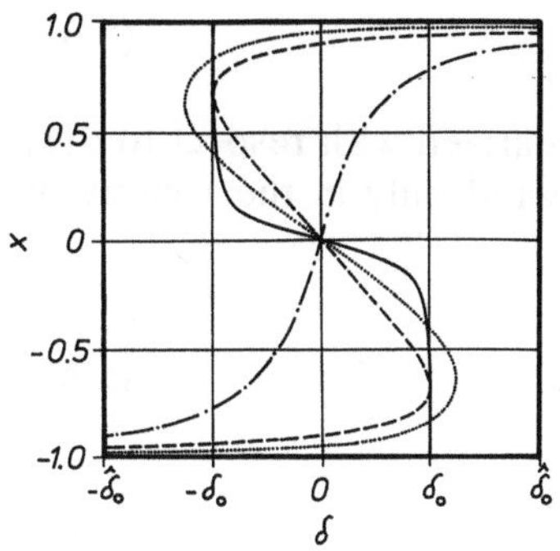

(———) $F_2(\delta, x) = 0$

(–·–·–) a) $F_1(\delta, x) = 0$ for small κ (one singular point – the origin)

(– – –) b) $F_1(\delta, x) = 0$ in the marginal case (3 singular points)

(· · · · ·) c) $F_1(\delta, x) = 0$ for very large κ (5 singular points)

Fig. 5.2. Typical graphs of the transcendental functions $F_1(\delta, x) = 0$ and $F_2(\delta, x) = 0$

The marginal Case b) with three singular points is characterized by the fact that not only (5.60, 61) hold good at the singular points $P_+(\delta_+, x_+)$ and $P_-(\delta_-, x_-)$ in the second and fourth quadrants, but also that the derivatives of the graphs $F_1(\delta, x) = 0$ and $F_2(\delta, x) = 0$ agree at P_+ and P_-; thus the condition

$$\left(\frac{dx}{d\delta}\right)^{(F_1)}_{P_\pm} = \left(\frac{dx}{d\delta}\right)^{(F_2)}_{P_\pm} \tag{5.62}$$

where $(dx/d\delta)^{(F_1)}$ and $(dx/d\delta)^{(F_2)}$ are the derivatives taken along the curves $F_1(\delta, x) = 0$ and $F_2(\delta, x) = 0$, is also satisfied. A straight-forward evaluation of (5.62) leads to the equivalent implicit condition

$$\kappa = \delta_0\beta\left[1 - \left(\frac{\bar{\delta}}{\delta_0}\right)^2\right] + \frac{1}{(1 - \bar{x}^2)} \tag{5.63}$$

which additionally has to be fulfilled by the parameters κ, δ_0 and β for this marginal case. In (5.63) the solutions δ_+, x_+ or $\delta_- = -\delta_+, x_- = -x_+$ of (5.60, 61) have to be inserted for $\bar{\delta}$ and $\bar{x}$ with $\delta_1 \equiv 0$.

The two-dimensional domain of parameters κ, δ_0 and β satisfying (5.63) for which there exist *three* singular points in the δ-x plane will be denoted by $\mathfrak{D}_2^{(3)}$. The two dimensional surface $\mathfrak{D}_2^{(3)}$ separates the three dimensional domain $\mathfrak{D}_3^{(5)}$ of the parameters κ, δ_0, β, for which *five* singular points exist, from the three dimensional domain $\mathfrak{D}_3^{(1)}$ for which only *one* singular point (namely the origin $\bar{\delta} = 0$, $\bar{x} = 0$) exists.

5.4.2 Stability Analysis

Secondly, it is relevant for discovering the type of paths traversed by $x(\tau)$ and $\delta(\tau)$ to know their behaviour in the vicinity of the singular points. Therefore a *linear stability analysis* will be performed in order to see whether the singular points are *stable* or *unstable* stationary points. With this objective the behaviour with time of the solutions in a small neighbourhood of a singular point $\bar{P}(\bar{\delta}, \bar{x})$ is investigated by introducing deviations $\eta(\tau)$ and $\xi(\tau)$ from $\bar{\delta}$ and $\bar{x}$:

$$x(\tau) = \bar{x} + \xi(\tau) \quad \text{and} \quad \delta(t) = \bar{\delta} + \eta(\tau) . \tag{5.64}$$

The equations of motion (5.53, 54) can now be linearized with respect to $\xi(\tau)$ and $\eta(\tau)$. This leads to the following, which are valid only in the vicinity of $\bar{P}(\bar{\delta}, \bar{x})$:

$$\frac{d\xi(\tau)}{d\tau} = a_1 \xi + b_1 \eta \tag{5.65}$$

$$\frac{d\eta(\tau)}{d\tau} = - a_2 \xi - b_2 \eta \tag{5.66}$$

with the coefficients

$$\left.\begin{aligned} a_1 &= \cosh^{-1}(\bar{\delta} + \kappa\bar{x})\,[\kappa - \cosh^2(\bar{\delta} + \kappa\bar{x})] \\ a_2 &= \gamma\beta\delta_0 \cosh^{-1}(\beta\bar{x}) \\ b_1 &= \cosh^{-1}(\bar{\delta} + \kappa\bar{x}) \\ b_2 &= \gamma\cosh(\beta\bar{x}) \end{aligned}\right\} \tag{5.67}$$

where $a_2, b_1, b_2 > 0$ and a_1 is unrestricted. According to standard methods (5.65, 66) can be solved by using

$$\left.\begin{aligned} \xi(\tau) &= \xi(0)\exp(\lambda\tau) \\ \eta(\tau) &= \eta(0)\exp(\lambda\tau) \end{aligned}\right\} \tag{5.68}$$

where the eigenvalues λ have to fulfill the determinant condition

$$\begin{vmatrix} a_1 - \lambda, & b_1 \\ - a_2, & - b_2 - \lambda \end{vmatrix} = (a_1 - \lambda)(- b_2 - \lambda) + a_2 b_1 = 0 . \tag{5.69}$$

This quadratic equation yields the eigenvalues

$$\lambda_{1,2} = 1/2 \left[(a_1 - b_2) \pm \sqrt{(a_1 + b_2)^2 - 4 a_2 b_1}\right] . \tag{5.70}$$

The singular point is a *stable focus* if the eigenvalues λ_1 and λ_2 both have negative real parts since then both $\xi(\tau)$ and $\eta(\tau)$ approach zero with increase in time. On the other hand, the singular point is *unstable* if the real part of at least one eigenvalue is positive.

Of great interest is the stability of the singular point $\bar{P}_0(\bar{\delta}_0 = 0,\ \bar{x}_0 = 0)$ which is, however, slightly shifted from the origin for a non-vanishing strategic choice bias δ_1. With $\delta_1 = 0$, $\bar{\delta} = 0$ and $\bar{x} = 0$ the parameters a_1, a_2, b_1 and b_2 assume the simple forms:

$$\left.\begin{aligned} a_1 &= \kappa - 1 \\ a_2 &= \gamma \beta \delta_0 \\ b_1 &= 1 \\ b_2 &= \gamma \end{aligned}\right\} \tag{5.71}$$

which leads to the eigenvalues

$$\lambda_{1,2} = 1/2 \left[(\kappa - 1 - \gamma) \pm \sqrt{(\kappa - 1 + \gamma)^2 - 4\gamma\beta\delta_0}\right]. \tag{5.72}$$

It is reasonable – and confirmed by the calculations presented in Sect. 5.5 – to assume relative large values for the trend reversal parameter β and the strategic choice amplitude δ_0. Thus

$$(\kappa - 1 + \gamma)^2 < 4\gamma\beta\delta_0 \tag{5.73}$$

can be assumed. In this case the eigenvalues (5.72) become conjugate complex and can be written as

$$\lambda_{1,2} = 1/2 \left[(\kappa - 1 - \gamma) \pm \mathrm{i}\sqrt{4\gamma\beta\delta_0 - (\kappa - 1 + \gamma)^2}\right]. \tag{5.74}$$

From (5.74) it follows immediately that there exists a *stable focus* at the origin of the δ-x plane under the condition

$$\kappa - 1 - \gamma < 0 \qquad \text{or} \quad \kappa < 1 + \gamma \tag{5.75}$$

and that the origin becomes an *unstable focus* if the condition

$$\kappa - 1 - \gamma > 0 \qquad \text{or} \quad \kappa > 1 + \gamma \tag{5.76}$$

is fulfilled.

5.4.3 The Limit Cycle Existence Theorem

Finally, the solutions of the equations of motion (5.53, 54) have to be investigated. A complete survey of all the possible types of solution to the equations in dependence on the parameters κ, δ_0, δ_1, β and γ will not, however, be made. The immediate aim is to derive *sufficient conditions* for those parameters which lead to the existence of an asymptotically periodic type of solution; this is the type of solution which has been anticipated qualitatively in terms of the investment cycle and in which the main interest lies. In other words: It is necessary for the substantial aspects of the model being considered to derive an *existence theorem* for *solutions approaching a "limit cycle"* under certain conditions of the parameters κ, δ_0, δ_1, β and γ.

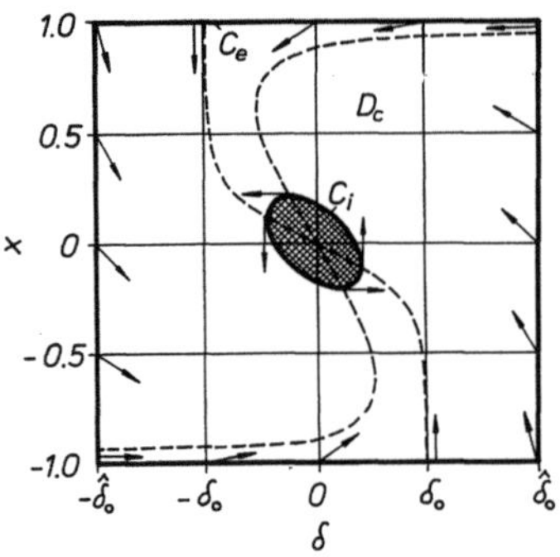

Fig. 5.3. Domain $\mathfrak{D}_c$ with interior boundary C_i and exterior boundary C_e. All trajectories enter $\mathfrak{D}_c$ and remain in $\mathfrak{D}_c$ as the fluxlines at the boundaries show

A limit cycle $C(t)$ is defined as a closed trajectory, i.e. a periodic solution to the equations of motion, with the property that there exists a domain $\mathfrak{D}_c$ around $C(t)$ so that all trajectories starting within $\mathfrak{D}_c$ approach $C(t)$ as $t \to \infty$. $\mathfrak{D}_c$ can be denoted as the "domain of attraction" and $C(t)$ as an "attractor". A special case of an attractor is an "infinitesimally small limit cycle", i.e. a stable focus.

a) The Poincaré-Bendixon Theorem. Before the existence theorem is stated, the famous *Poincaré-Bendixon theorem* in a – in context – relevant version should be formulated:
Assumptions: Consider two autonomous first order differential equations for the variables $\delta(\tau)$ and $x(\tau)$ and suppose that a finite domain $\mathfrak{D}$ exists in the δ-x plane such that:

a) no singular points are situated in $\mathfrak{D}$.
b) all trajectories $\delta(\tau)$, $x(\tau)$ starting inside or on the boundary of $\mathfrak{D}$ at time $\tau = 0$ remain in $\mathfrak{D}$ for $0 < \tau < \infty$.

Theorem: In this case there must exist a (at least one) limit cycle within $\mathfrak{D}$ and all trajectories in $\mathfrak{D}$ either *are*, or *approach* as $\tau \to \infty$, a limit cycle.

b) Statement and Proof of the Existence Theorem. The existence theorem needed can now be stated and proved.

Assumptions: Suppose that the parameters κ, δ_0, β, γ, $\delta_1 = 0$ satisfy the conditions

$$\left.\begin{array}{l} 1)\ \{\kappa, \delta_0, \beta\} \in \mathfrak{D}_3^{(1)}\,, \\ \quad \text{i.e. the origin is the only singular point} \\ 2)\ (\kappa - 1 + \gamma)^2 < 4\gamma\beta\delta_0 \\ 3)\ \kappa > 1 + \gamma\,. \end{array}\right\} \tag{5.77}$$

These imply that the eigenvalues λ_1 and λ_2 belonging to $\overline{P}(\overline{\delta} = 0, \overline{x} = 0)$ are conjugate complex with a positive real part and so the one singular point $\overline{P}(0, 0)$ is an *unstable focus*.

Theorem: The domain $\mathfrak{D}_0$ of the δ-x plane bounded internally by an infinitesimally small ellipse enclosing the origin and externally by the straight lines $\delta = \delta_0$, $x = 1$, $\delta = -\delta_0$ and $x = -1$ (Fig. 5.3) fulfills the premise of the Poincaré-Bendixon theorem that:

a) there are no singular points situated in $\mathfrak{D}_c$,
b) all trajectories starting from the boundary of $\mathfrak{D}_c$ enter $\mathfrak{D}_c$ and remain in $\mathfrak{D}_c$.

Proof: Premise a) follows from condition 1) of (5.77), according to which the origin is the *only singular* point. The origin, however, is not an element of $\mathfrak{D}_c$.
Premise b) follows from the fact that according to conditions 2) and 3) of (5.77) the origin is an *unstable focus* with all trajectories spiraling out of it and therefore *entering* $\mathfrak{D}_c$ from its interior boundary C_i; also on the exterior boundary C_e:

$$\left.\begin{aligned} \frac{dx}{d\tau} &= [\tanh(\delta + \kappa) - 1]\cosh(\delta + \kappa) < 0 \qquad \text{for } x = 1 \\ \frac{dx}{d\tau} &= [\tanh(\delta - \kappa) + 1]\cosh(\delta - \kappa) > 0 \qquad \text{for } x = -1 \end{aligned}\right\} \tag{5.78}$$

$$\left.\begin{aligned} \frac{d\delta}{d\tau} &= -\gamma[\delta_0 \tanh(\beta x) + \delta_0]\cosh(\beta x) < 0 \qquad \text{for } \delta = \delta_0 \\ \frac{d\delta}{d\tau} &= -\gamma[\delta_0 \tanh(\beta x) - \delta_0]\cosh(\beta x) > 0 \qquad \text{for } \delta = -\delta_0 \end{aligned}\right\} \tag{5.79}$$

and (5.78, 79) show that all trajectories crossing C_e are *directed inward* into $\mathfrak{D}_c$.

Conclusion: From the Poincaré-Bendixon theorem it now follows that within $\mathfrak{D}_c$ a limit cycle, i.e. a periodic solution of the equations of motion $C(t)$, must exist and that all other solutions starting from any point within $\mathfrak{D}_c$ approach this limit cycle. $\mathfrak{D}_c$ is a domain of attraction for $C(t)$.

Remark: The assumptions 1–3) of (5.77) are *sufficient* but *not necessary* conditions for the existence of a limit cycle. This means that there could also exist limit cycle solutions under more general conditions for the parameters κ, δ_0, δ_1, β and γ. On the other hand, it will be seen that initially quasi-cyclic trajectories exist which break down into a stable solution after some oscillations (Figs. 5.9 a and b).

If the assumption 3) of the existence theorem is changed into

$$\kappa < 1 + \gamma \, , \tag{5.80}$$

the case is obtained for which the origin is the only singular point and is a *stable focus*. Defining the domain $\mathfrak{D}_0$ as the *full rectangle* with the same exterior boundary as $\mathfrak{D}_c$ but with no internal boundary, using (5.78, 79) $\mathfrak{D}_0$ is now a domain of attraction for the stable focus $\overline{P}(0, 0)$, i.e. with (5.80) instead of (5.76) all the trajectories in $\mathfrak{D}_0$ spiral into the stable solution $\overline{P}(0, 0)$. This behaviour is confirmed by the numerical solutions of the model presented in Sect. 5.5 (Figs. 5.4, 5). The main importance of these two different types of solution lies in the implied proof that the answer to the question as to whether an economy tends towards a stationary state or undergoes non-equilibrium oscillations depends in a sensitive way on the decision psychology of the entrepreneurs as expressed by the decision psychology parameters κ, δ_0, δ_1, β and γ. In particular there exist "critical values" of these parameters. Crossing these critical thresholds elicits a transition from one into the other type of solution. (See the transition from Figs. 5.4a and b with $\kappa = 1.0$ to Figs. 5.6a and b with $\kappa = 1.6$, which crosses the marginal case of Figs. 5.5a and b which have a critical threshold value of the coordinator $\kappa_c = 1.5$).

5.5 Numerical Analysis Based on the Model

In this Section explicit solutions to the equations of motion, (5.53, 54), are presented for some important combinations of parameters. Firstly, in Sect. 5.5.1 numerical experiments are conducted in order to explore the model's dynamic behaviour for a hypothetical economy. Secondly, in Sect. 5.5.2 the model is employed in an econometric analysis of changes in volume and composition of industrial investment in a real economy, that of the Federal Republic of Germany.

5.5.1 Model Solutions for the Motion of a Hypothetical Economy in Ideal Time

The computer assisted experiments reported in this section serve the purpose of numerically substantiating the analytical results presented in Sect. 5.4 and of demonstrating phenomenologically that different combinations of parameters lead to different types of solutions, some representing a motion of the economy approaching a stationary state, and others representing a non-equilibrium motion with periodic limit cycle behaviour.

In depicting the results use is made of two types of plots, the so called a-plot and b-plot. In terms of a-plots six solutions for $\delta(\tau)$ and $x(\tau)$ are presented as trajectories in the δ-x plane in Figs. 5.4a–9a. These figures also contain the functions $F_1(\delta, \overline{x}) = 0$ and $F_2(\delta, x) = 0$ which are the geometrical loci of the horizontal and vertical fluxlines, the intersections of which represent the singular points $\overline{P}(\overline{\delta}, \overline{x})$ for the respective combinations of parameters. Additionally, as b-plots, Figs. 5.4b–9b show the paths of $\delta(\tau)$ and $x(\tau)$ with scaled time $\tau = 2\nu t$ for the same parameter combinations.

Experiment 1. To begin with, a parameter combination is chosen for which the origin $\delta = 0$, $x = 0$ is the only singular point and a stable focus of the motion; all trajectories approach the stationary solution $\delta = 0$, $x = 0$. Thus the

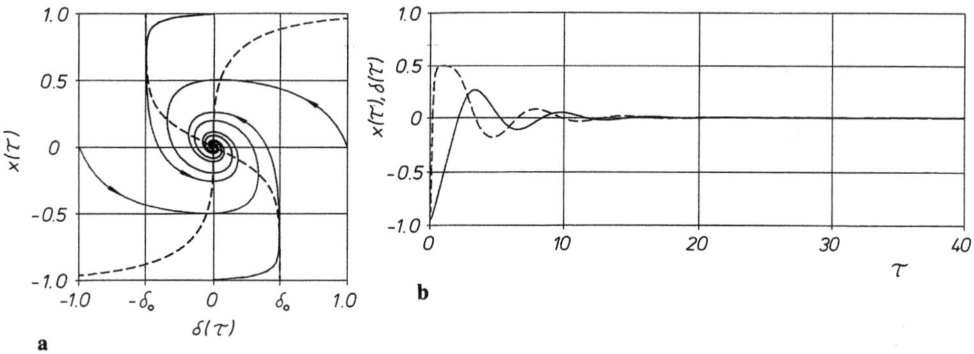

Fig. 5.4. **(a)** Trajectories in the δ-x plane approaching the one stable focus at the origin $\overline{P}(0, 0)$ for $\kappa = 1.0$, $\delta_0 = 0.5$, $\delta_1 = 0$, $\beta = 4.0$ and $\gamma = 0.5$. (—) Trajectories; (---) $F_{1,2}(\delta, x) = 0$. **(b)** Paths of $x(\tau)$ (—) and $\delta(\tau)$ (---) for the parameters of part (a)

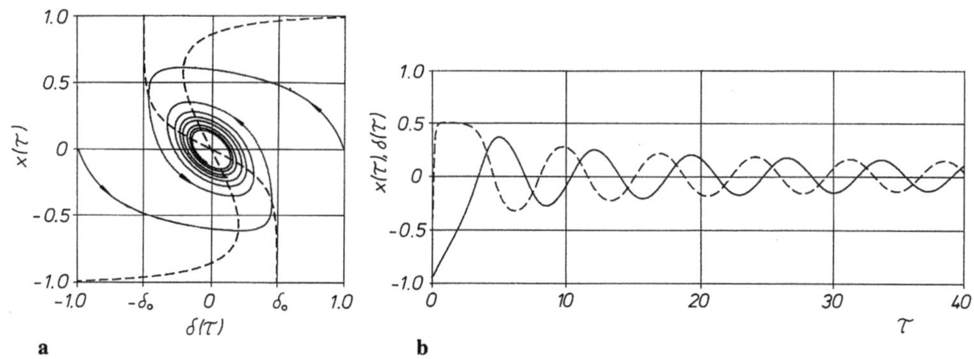

Fig. 5.5. **(a)** Trajectories in the δ-x plane. The approach of the one marginally stable focus at the origin $\overline{P}(0, 0)$ is critically slowed down when $\kappa = 1.5$, $\delta_0 = 0.5$, $\delta_1 = 0$, $\beta = 4.0$ and $\gamma = 0.5$. (—) Trajectories; (---) $F_{1,2}(\delta, x) = 0$. **(b)** Path of $x(\tau)$ (—) and $\delta(\tau)$ (---) for the parameters of part (a)

damping out of cyclic behaviour, a phenomenon that troubles disequilibrium models in situations of zero-growth, is but a special case among many others in the non-equilibrium model being discussed.

Experiment 2. At the threshold value $\kappa = 1.5$ to the transition to another type of solution, namely that of solutions with sustained cycles, there occurs a critical reduction of the damping effect. Although the origin is still the only singular point and is still – marginally – stable, the oscillatory relaxation takes much more time than in the previous case.

Experiment 3. Sustained cycles occur for a parameter combination that satisfies the assumptions of the limit cycle existence theorem. Figure 5.6a exhibits the limit cycle (see also Fig. 5.6b). Again, the origin is the only singular point which, however, has now become an unstable focus of the motion of the economy.

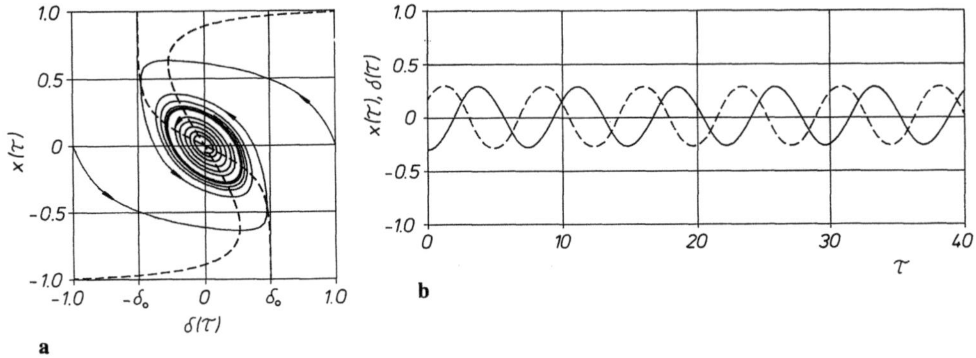

Fig. 5.6. (a) Trajectories in the δ-x plane for one unstable focus at $\overline{P}(0, 0)$ and one limit cycle for $\kappa = 1.6$, $\delta_0 = 0.5$, $\delta_1 = 0$, $\beta = 4.0$ and $\gamma = 0.5$. These fulfill the assumptions (5.77) of the existence theorem. (—) Trajectories; (---) $F_{1,2}(\delta, x) = 0$. **(b)** Paths of $x(\tau)$ (—) and $\delta(\tau)$ (---) for the parameters of part (a). The motion is periodic and traverses the limit cycle

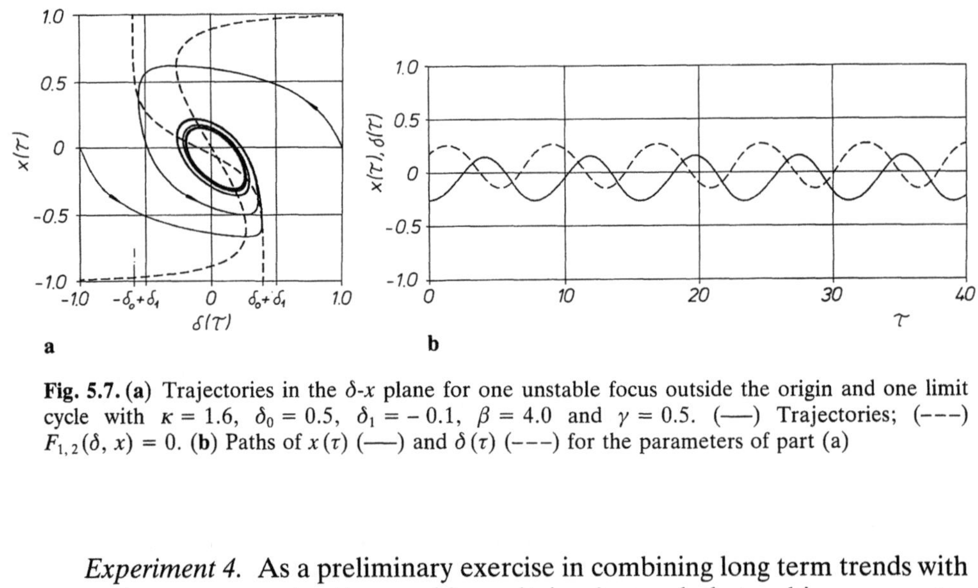

Fig. 5.7. (a) Trajectories in the δ-x plane for one unstable focus outside the origin and one limit cycle with $\kappa = 1.6$, $\delta_0 = 0.5$, $\delta_1 = -0.1$, $\beta = 4.0$ and $\gamma = 0.5$. (—) Trajectories; (---) $F_{1,2}(\delta, x) = 0$. **(b)** Paths of $x(\tau)$ (—) and $\delta(\tau)$ (---) for the parameters of part (a)

Experiment 4. As a preliminary exercise in combining long term trends with expansionary or contractionary biases it is observed that a bias parameter $\delta_1 \neq 0$ can be introduced so that the general type of the solution (limit cycle) still exists, but the form of the limit cycle is altered; the periodic motion is distorted (higher maxima and lower minima), as shown in Figs. 5.7 a and b.

Experiment 5. Over and above longitudinal influences on the form of the limit cycle, cross-sectional influences can occur, such as an influence on the frequency of one nation's cycle by the business activities in other nations (locomotion theory). The resulting strategic reactions of this one nation's investors may further distort the limit cycle, so that a modified strongly non-sinusoidal reaction pattern emerges; this then favours or disfavours a rush into certain types of investment; see Figs. 5.8 a, b.

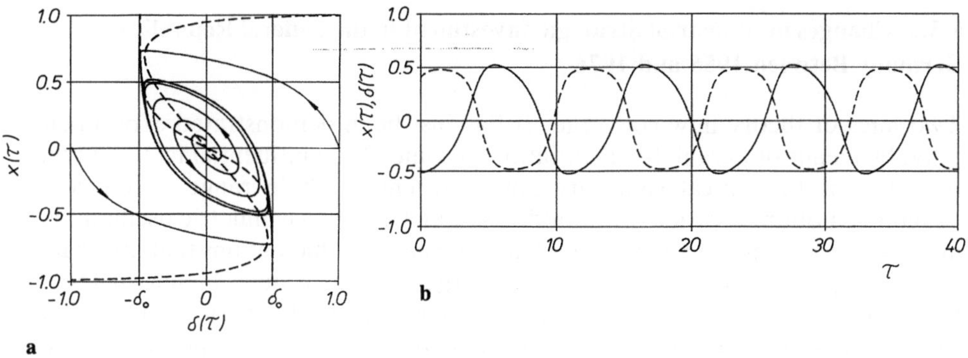

Fig. 5.8. (**a**) Trajectories in the δ-x plane for one unstable focus at $\overline{P}(0, 0)$ together with a limit cycle for $\kappa = 1.9$, $\delta_0 = 0.5$, $\delta_1 = 0$, $\beta = 4.0$ and $\gamma = 0.5$; these still fulfill the assumptions (5.77) of the existence theorem. (—) Trajectories; (---) $F_{1,2}(\delta, x) \doteq 0$. (**b**) Paths of $x(\tau)$ (—) and $\delta(\tau)$ (---) for the parameters of part (a). The motion is periodical and non-sinusoidal

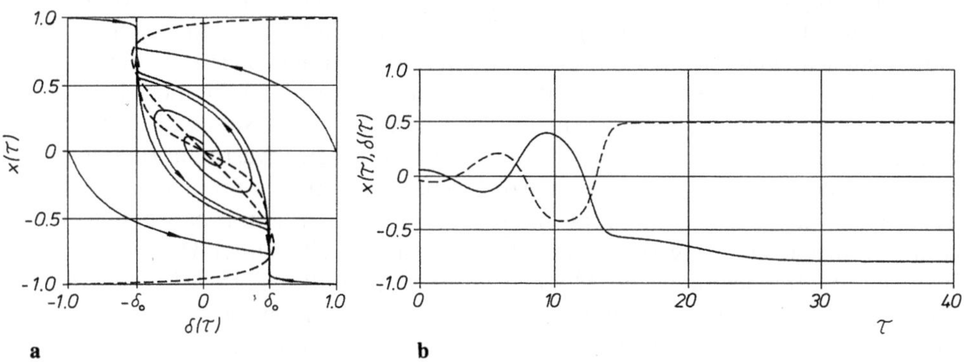

Fig. 5.9. (**a**) Trajectories in the δ-x plane approaching one of five singular points with $\kappa = 2.0$, $\delta_0 = 0.5$, $\delta_1 = 0$, $\beta = 4.0$ and $\gamma = 0.5$; there exists no limit cycle. (—) Trajectories; (---) $F_{1,2}(\delta, x) = 0$. (**b**) Paths of $x(\tau)$ (—) and $\delta(\tau)$ (---) for the parameters of part (a). After breakdown of quasi-periodic motion stationary values $\bar{x} \approx -0.8$ and $\overline{\delta} \approx 0.5$ are reached

Experiment 6. Lastly, yet another type of solution emerges when but minor parameter changes are made. These lead, however, to a structural change in the pattern of singular points. With the parameters of Fig. 5.9 a, for example, there occur five singular points; two of these are stable attractors for a phase transition, upon which the formerly quasi-periodic motion breaks down and quickly ends in a stationary solution (such as long term temporary over- or underemployment equilibra [5.21]. Here, the stationary solution is reached with $\overline{\delta} \approx 0.5$ and $\overline{x} \approx -0.8$ (Fig. 5.9 b).

5.5.2 Changes in Industrial Strategic Investment in the Federal Republic of Germany Between 1956 and 1978

Two lines of theory now converge. It has just been demonstrated experimentally, by means of parametrical variation of some of the influential factors in the model, that theoretical solutions can represent a whole range of types of economic motion; a variety of periodic, symmetric and asymmetric oscillations, and also some vanishing, accelerated and decelerated fluctuations that approach an existing mean value trend line or diverge towards a new, higher or lower trend line have been demonstrated. And it is exactly this variety of possible types of motion which has to be explained when theory is compared with the variety of economic motions observable empirically. It has therefore been shown that by incorporating the micro-economic strategic investment decisions of entrepreneurs into the model macro-economical fluctuations can be interpreted and that the influence of macro-economic change on the micro-economic level of entrepreneurship can be explained.

In devising a test for the model with data for a real economy, two restrictions apply. The first restriction is *spatial*: The non-availability of time series on industrial strategic investments (expansionary and rationalizing) in some countries narrows down the choice of national economies to which the model can be readily applied. But even in countries such as the United States of America, where data are available but only for one type of strategic investment (expansionary) and not for the other rationalizing type (the latter is reported together with tactical investments or replacement), some relevant empirical regularities could be predicted using the model. For example, "The Other Half of Gross Investment: Replacement and Modernization Expenditures" [5.22] varies over time, while replacement itself may occur as a constant proportion of capital stock (gross fixed assets) as has been observed by *Jorgenson* [5.23] and *Jorgenson* and *Stephanson* [5.24]. On the other hand, the fluctuations in replacement and modernization expenditures are less pronounced than the fluctuations in expansionary investment, which according to *Eisner* [Ref. 5.25, p. 181] explain almost 80% of the variation in annual capital expenditures in the United States' business sector. Consequently, it can be predicted using statistical methods that the explanatory power of an investment analysis falls under the 80% line suggested by Eisner if the analysis mixes expansionary investment with investment for modernization and replacement since the analysis of "strategic" investment would be "contaminated" with some "tactical" investment, namely for replacements. On the other hand, the explanatory power of an analysis of industrial investment in plant and equipment in the United States would probably exceed 80% if the analysis were restricted to "strategic" investment proper, namely expansionary and rationalizing investment (the latter now being buried under "modernization and replacement"). The availability of pertinent data is one of the reasons for the choice of the Federal Republic of Germany as the test unit for analysis.

The second restriction for the application of the model as it stands is *temporal*: Times of discontinuity have been excluded from the test as these represent changes in economic regime, or phase transitions. Although such transitions can be handled by the model as the sixth experiment in the previous subsection shows (Fig. 5.9 a, b), additional theory (long term economic theory) would have to be incorporated – for instance in the form of an equation of motion for the investors' interaction parameter – in order to handle discontinuities systematically. It is well known that differential equation models of only three coupled dynamic variables (e.g. the Lorenz model) already possess "strange attractor" solutions side by side with those of the limit cycle type. Though fully deterministic, such solutions could represent quasi-chaotic economic motion. These questions of long term economic theory and model architecture and how they relate to possible phase transitions in times of discontinuity, will not be further discussed here.

For the period 1955–1980 data are available on expansionary and rationalizing industrial investment in the Federal Republic and moreover this period appears to be void of a major discontinuity. Nevertheless this period of a quarter of a century certainly does not build a continuum; in terms of economic policy and government regulation it can be divided into two main subperiods each of about eleven years whereby the first four or five years of the second subperiod can clearly be designated as interventionistic. Investment activity was therefore under the influence of the "regulator" δ_1 to varying degrees, which can be assumed as changing for the three intervals 1955–1965, 1967–1971 and 1973–1980. Similarly the general loss of purchasing power is reflected in an increasing "inflator" g. Furthermore the technological base of the german industry changed and matured during these years in more or less the way suggested and modelled by *Utterback* and *Abernathy* [5.26] as well as several other authors [5.14, 21]. According to their findings, industries develop in terms of technology and market structure by going through stages which depend on the age of the technology and other related factors in the particular industry concerned, but for which the following regularity holds for the post-war growth industries in most western countries: The 1950s were dominated by product innovation in diversifying industry, the 1960s by process innovation in concentrating industry and the 1970s by pseudo-innovations in nearly all industries, just as the 1980s are expected to be dominated by a cluster of basic innovations in some key industries. It can therefore with justification be thought that these patterns in the rate and direction of innovation have created, and have been the consequence of, a high degree of conformity in strategic behaviour (namely, in the timing of innovation, synchronization of investment and imitation of business tactics) in the first subperiod under consideration. In the second subperiod, on the contrary, there appears to have been a low degree of conformity. This alteration is reflected in the "high", "high" and "low" values of the "coordinator" κ chosen in modelling the three intervals 1955–65, 1967–71 and 1973–80. Similarly, in the three time intervals the value of the "accelerator" ν was chosen to be "small", "medium" and "large" because it reflects the shortening of successive product life cycles as major industries advanced out of a phase of large improvement effects (true product innovation) to minor improvement effects (mere pseudo-innovation) which do not last very long.

In Table 5.1 the chosen values of the parameters are listed for the reader who may wish to replicate the empirical analysis that follows and the investment data for the Federal Republic of Germany are given in Table 5.2.

Table 5.1. Model Parameters for the Economy of the Federal Republic of Germany

Period	γ	β	δ_0	δ_1	g	κ	ν
1955–1965	0.13	10	0.5	−0.3	0.2	1.5	0.145
1967–1971	0.13	10	0.5	+0.3	0.5	1.5	0.225
1973–1980	0.13	10	0.5	−0.3	0.7	0.7	0.400

The original data, collected by a german institute for economic research (The IFO-Institute in Munich) by means of questionnaires sent to a (representative) sample of german industrial corporations, has been econometrically corrected for a stronger rationalization bias [5.21] so that the regression line of Fig. 5.10 a depicting the data possesses a smaller intercept and a steeper descent than it would have in the original data version, which incorporates an expansionary bias. Figure 5.10 b shows the investment structure data $z(t)$ which is the investment structure index $Z(t)$ cleaned of the linear trend. *Weidlich* and *Haag* [5.27] have shown that the trend deviation pattern is robust, i.e. insensitive to the form of the trend function. Also, since the investment structure index without trend is to be used, the difference in built-in biases between the original data version and the Mensch version should have no effect on the matching of the observed trend deviations with the theoretical values calculated from the model.

Figure 5.11 shows the result of the empirical analysis which consisted of a piecewise application of the model to the set of parameters specified in Table 5.1 for the three intervals 1955–1965, 1967–1971 and 1973–1980.

From the mid-fifties through the mid-sixties the matching external conditions chosen correspond to those implying the existence of a limit cycle as indicated by the limit cycle existence theorem. During these years the Federal Republic was still enjoying its "economic miracle" which was nurtured by both post-war reconstruction of plant, equipment and infrastructure and by a high rate of product innovation in a number of growing international industries. A strong expansionary bias and bias reinforcement ($\kappa = 1.5$), however, lead to a strongly non-sinusoidal distortion of the periodic motion of the investment structure index; again, as in the case of the United States, the fluctuation in expansionary investment explains most of the variation in the economic indicators of aggregate change. Even in the expansionary phase industrial investors changed quickly to rationalizing investment after a relatively short expansionary boom and they changed only rather slowly back again in the direction of expansionary investment even after a relatively long period of rationalization. Thus the effects of entrepreneurial decision processes as reflected by the path of $z(t)$ in Fig. 5.11 reveal a persistent and perhaps growing concern for rationalization and productivity advancement even in the interval 1955–65 when the economy expanded at a high rate. On the other hand, preparation for the "quantum leaps" which then took place, takes time. The development of new or better

Table 5.2. Investment Data for the Federal Republic of Germany 1956–78

Year	R-Investment [10^9 DM]	E-Investment [10^9 DM]	$Z = \frac{E - R}{E + R}$
1956	16.4	11.4	−0.183
1957	15.9	10.8	−0.197
1958	16.6	8.2	−0.342
1959	17.7	10.2	−0.274
1960	22.2	12.6	−0.282
1961	24.2	13.6	−0.282
1962	24.2	11.2	−0.369
1963	22.4	10.6	−0.360
1964	23.5	11.2	−0.360
1965	26.1	13.0	−0.342
1966	25.9	11.6	−0.388
1967	23.0	7.2	−0.527
1968	22.3	10.4	−0.369
1969	30.0	21.6	−0.162
1970	34.9	23.2	−0.205
1971	34.4	18.6	−0.298
1972	31.3	12.4	−0.439
1973	30.5	13.8	−0.379
1974	28.5	10.2	−0.471
1975	25.9	6.2	−0.626
1976	26.5	6.8	−0.626
1977	27.5	6.6	−0.613
1978	28.9	7.0	−0.613

products, and the planning of new or bigger plants to produce them, must reach back over the years preceeding periods of growth in the business cycle.

During the interval 1955–65 the strategic choice parameter $\delta(t)$, the "alternator", stays only briefly in the rationalization mode and switches quickly back into the expansionary mode as soon as rationalization investment strategy has taken effect. From the mid-sixties onwards, however, the alternator took a different course, staying mostly in the rationalization mode only rarely and briefly switching over to the expansionary mode. The drop in the propensity to finance extended plant and equipment as well as the research and development work that leads to growing business, is seen to occur as early as 1966 or 1967. As process innovation was the gist of the rationalization atmosphere that prevailed from about 1965 on, a confirmation has been obtained using german statistics for the Utterback-Abernathy model of product and process innovation, which indicates a fairly general switch-over from product innovation to process innovation at about that time. Furthermore, and as has been already suggested by *Mensch* [5.14], the seventies were a period of relatively weak propensity to innovate both products and processing in comparison to the previous years, and the time-path of the alternator found to fit the observed investment behaviour

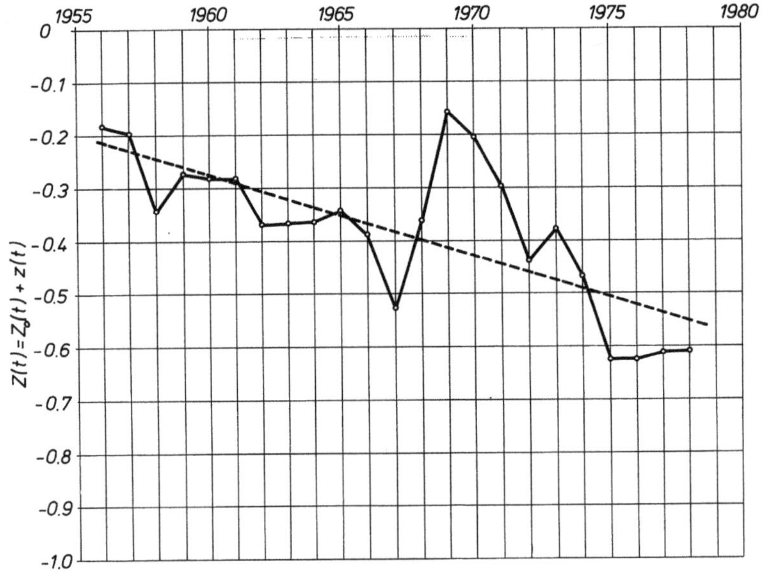

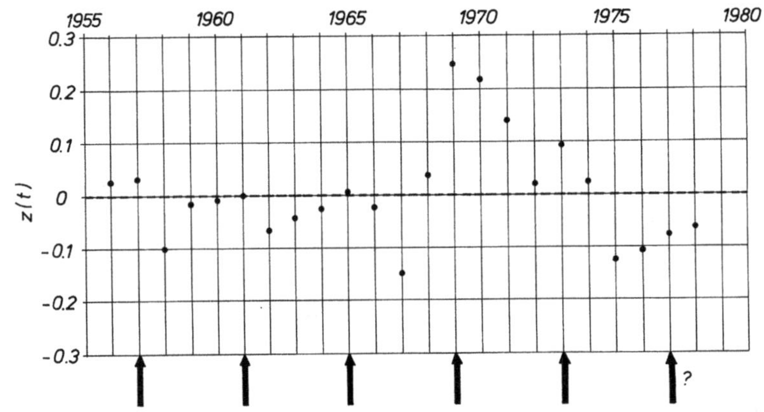

Fig. 5.10. (a) Investment structure index $Z(t) = Z_0(t) + z(t)$ in the Federal Republic of Germany, 1955–1978 (data by IFO). A linear time dependence of the long term component $Z_0(t)$ (---) is assumed. **(b)** Short term investment structure index $z(t)$ in the Federal Republic of Germany, 1955–1978. The long term component $Z_0(t)$ of part (a) has been subtracted

does indeed portray the predominance of essentially pseudo-innovation during the 1970s.

During the late sixties after the neo-liberal incumbant government fell on the issue of a mere 3% inflation rate, the newly formed "Grand Coalition" in the Federal Republic of Germany resorted to a massive expansionary policy programme. Its effect on the investment structure index is clearly visible as an all-time peak in 1969/1970 called the "Schiller Effect" after Schiller, the accomplished Keynesian economist, who became both minister of finance and of

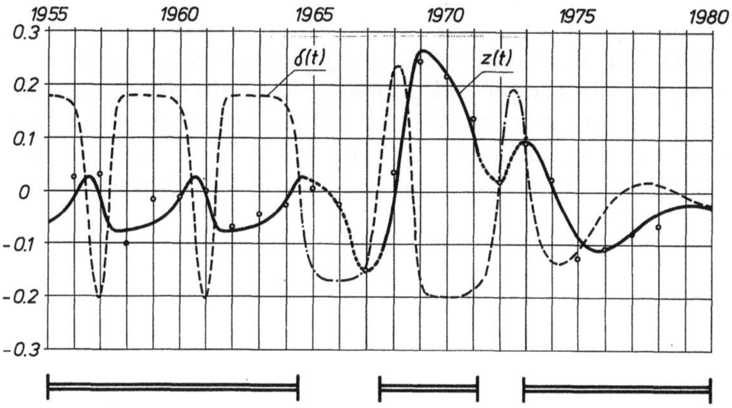

Fig. 5.11. Path of the short term component $z(t)$ of the investment structure index (—) and of the alternator $\delta(t)$ (---) for appropriate choices of trend parameters in comparison with the observed investment structure data

economic affairs. This irregularity in economic conditions has been adjusted for by temporarily boosting the "regulator" δ_1. The very fact that after the Schiller period the regulator had to be taken back to its initial level in order to achieve the fit depicted in Fig. 5.11 for the period after 1972 may teach more than that it is simply necessary to alter it in order to model unusual periods. It may lend further evidence to the not uncommon view that forceful demand management only forces industrial firms to consider more quickly the supply side limits to growth.

Although some economists argue that the oil shock caused the "slumpflation" (M. Freedman) after the hyperboom around 1970, an evolutionary explanation may be more appropriate. For the interval after 1971 it was not possible to fit the observed motion of the investment structure index (net of trend) with model data on the basis of parameters that satisfy the conditions of the limit cycle existence theorem. The matching values of the "inflator" g, which could not be set lower with the given price developments, and of the "accelerator" ν, which reflects the shortened product life cycles of pseudo-innovations, fit together only by using a "coordinator" value κ below unity, which is in the realm of damped fluctuations. Marketing managers speak of "a fluttering cycle" in reference to this quasi-stationary state in which the low amplitude and high frequency of the branch cycles do not add up to either an upswing or a true downswing in macro-economic aggregate.

Obviously this is a state of high micro-economic friction and loss of energy, which carries already the seeds for a transition into another cyclic state in which the investors' interaction parameter κ would exceed the critical value κ_c. Needless to say, the economy need not necessarily resume its cyclic motion with an upswing from the current intermediate level of investment activities.

The model is not designed to predict the investment structure index. But it can be used for diagnostic purposes: If the data are meaningful and the model is reasonable, which the authors believe, then industry in the Federal Republic is in a state of de-phasing (asynchronization of micro-economic developments for the diverse firms, industrial branches and sectors). Thus, energetically, it is getting "ready" for a phase transition and preparing itself structurally for the breakthrough of a whole set of basic innovations. Whether this will in fact take place could only be answered by further investigations which encompasse many more factors than the short-term business cycle theory presented here.

6. The Interaction of Competitive Macrosocieties

Present world politics are dominated by the fact that (at least) two competitive macrosocieties have emerged in the course of recent historical development. These rivalling macrosocieties compete on almost all levels, in particular on the ideological, scientific, economic and military levels; military "progress" has reached the point that a capacity for the self-destruction of mankind has been achieved.

History shows us that in the majority of cases competing macrosocieties (states, or groups of states united by common interests and/or ideologies) have undergone phase transitions between three main phases: "peace", "tension" and "war". The present state of macrosociety relations – lying somewhere between peace and tension – could be denoted as "coexistence".

The threat to, or the disruption of "peace" towards one of the other phases is not removed by the fact that the overwhelming majority of individuals in all countries of the world fervently want the permanent stability associated with "peace". This sad conclusion comes from the observation that statesmen, even if doing their best to ensure peace[1], are not able to control all the complex interactions occuring between collective levels of individuals. In particular, the sensitive areas of the collective modes of political psychology and ideology as well as those of economic and military power, which constitute the core of relations between macrosocieties, can give rise to unpredictable and uncontrollable phase transitions.

6.1 Reconsideration of the Problem of Model Construction

If semi-quantitative sociology is to make any sense, it cannot evade the question as to whether it should contribute to the investigation of the great problems and threats of our era that arise from the existence of competing macrosocieties. The question leads back to the general problems discussed at the end of Chap. 1 and this discussion should now be resumed in the light of the other chapters and with respect to the current state of macrosociety relationships.

[1] The German authors, looking back at the history of their own country, know very well that this is not at all always the case!

Sociology, in particular semi-quantitative sociology, has an interesting property: It is simultaneously a *"metalevel"* and a *"sublevel"* of the society:

Being a metalevel, theoretical sociology describes and explains the interaction of material and ideological structures of the society.

Being a sublevel, sociology itself is a *part* of the society interacting with other parts and influencing their dynamics.

Hence, in a certain sense, theoretical sociology is *self-reflexive* since it looks at structures of the society, of which it is itself a part.[2]

It is the optimistic hope that among the conceivable levels of sociology there will slowly develop a *meta-ideological level of thinking* beyond the level of ideologies which are usually blind to their own dangerous consequences. The authors believe that only on the basis of sober insight into sociological structures, including ideologies and their dynamics, the global catastrophy of a new, and perhaps final phase transition to warfare has a chance of being avoided.

Semi-quantitative sociology seems to be particularly suitable for a meta-ideological approach in looking "sine ira et studio" at the effects of economic and ideological marcrovariables on the dynamics of societies.

The following question, however, has to be reconsidered: In which sense do the potential achievements of a *semi-quantitative sociology* go beyond those of a meta-ideological but traditional *qualitative sociology*?

The answer is: The semi-quantitative line of reasoning does not *substitute* but *presumes* and *includes* qualitative insight on a new level of argumentation which includes a *feedback* between *qualitative* and *quantitative* argumentation within the framework of model construction. On this level intuition and experience in the fields of both political psychology and mathematical structures join and reinforce each other in the following way:

1. *Qualitative insight* on relevant abstract (political, psychological, ideological) and material (economic, environmental, military) aspects of a society leads in its advanced form to an estimation of the *extent* and *intensity* of these factors. As a result, a qualitative model can be moulded of their interactions and of their influences on the dynamics of a society.

2. At such a point *quantitative argumentation* can be applied: The "qualities" and their "intensities" are *translated into grossvariables* of the socio-configuration and of the material situation. Finally, a dynamic model in terms of equations of motion for the grossvariables is tailored which takes the qualitative model mentioned above into account.

3. Once such a tentative quantified level has been reached, a *reciprocal procedure* sets in: The qualitative concepts are now reconsidered in the light of the achieved quantification. It may turn out, for instance, that the equations have to be modified after scrutinizing the dynamics implied by them, because they do not describe the reality correctly. It may also turn out that concepts seemingly reconcilable on a qualitative level prove to be incompatible and in need of revision on the more logically demanding quantitative level.

---

[2] The fundamental importance of self-reflexivity for the structure of formal logic has recently been stressed by *Hofstaeder* [6.1]

Let it be assumed that the application of this process of mutual modification between qualitative and quantitative argumentation finally leads to a quantitative model reflecting at least a part of the "truth"!

If such a model is *simple enough* to gain insight into the mathematical structure of its solutions and, on the other hand, *extensive enough* to include the modelling of sociological processes of a realistic complexity, it could lead to results that go further than purely qualitative argument.

Classifications of solutions, insight as to the influence and the criticality of trend parameters, as to the stability or lability of situations etc. could then be obtained which are beyond the capacity and certainty of a purely qualitative argument. Of course, the latter remains the basis of the quantification, but its implications can be investigated much more exhaustively by the quantitative model.

The decisive steps 2) and 3) of the translation into quantified grossvariables and the relation between quantitative and qualitative levels of argumentation will now be considered in more detail.

The grossvariables of political psychology are not available in a completely evident and explicit form through a direct method of measurement, as is often the case in physics or other "natural" sciences, so that relevant grossvariables usually have to be defined indirectly and implicitly. Two possible procedures in defining grossvariables for introduction into a model are:

a) The Quantification via the Socio-Configuration. Here one starts from the attitudes of the members of the society and finds a quantification of the intended grossvariable by taking an appropriate (perhaps) weighted average over the socio-configuration that agrees with the intuitive meaning of the grossvariable. The dynamics of the grossvariable are then in principle determined by the transition probabilities between attitudes. In Section 3.4 this quantification method including further approximations for obtaining closed dynamic equations for the grossvariables has been described, and in Section 6.2 the variable of "political psychology" will be introduced in this way.

b) The Quantification Using Global Properties of the Grossvariable. An explicit knowledge of the distribution of attitudes – i.e. the socio-configuration – is often not available. On the other hand, experienced politicians, sociologists and economists usually have a detailed understanding of the global properties of the intended grossvariables and of the driving forces determining their dynamics. This knowledge may be used *directly* in setting up equations of motion for the grossvariables, also taking into account their interaction.

Conversely, grossvariables are to some extent *implicitly defined* by the context of their equations of motion. Arguing anew about the qualitative meaning and the quantitative dynamic behaviour of coupled grossvariables on the basis of their equations builds the qualitative/quantitative feedback level of argumentation mentioned above.

6.2 The "Minimal Model"

In looking for global properties of the interaction of competing macrosocieties it is not the intention here to introduce variables describing "microscopic" details of this interaction. Instead a rather "coarse-grained" approach will be used (other authors [6.2–7] have proposed interesting and much more detailed models of world-wide *economic* interactions).

There is, however, one aspect in which the constructed model will go further (to the authors' knowledge) than hitherto proposed models: In the spirit of the concepts developed in the preceding chapters the effect of *political psychology* on the interaction of competing macrosocieties will be *quantitatively included.*

The minimal requirement of such a model is that it should describe – in a coarse-grained manner – the properties of the quasi-stationary states "peace" and "tension", the stability or lability of these states and the possibility of eventual transitions between them. Furthermore, the model cannot avoid including a consideration of the possibility of a transition to the state "war". Only in this way can an estimation be made as to which trend parameters in the model are responsible for an increase or decrease of the threat of this transition.

6.2.1 The Grossvariables of the Model

Let the two macrosocieties be denoted by $\mathcal{S}_1$ and $\mathcal{S}_2$. $\mathcal{S}_1$ and $\mathcal{S}_2$ are treated *symmetrically* in so far as the same set of grossvariables for each system is introduced and a symmetry of the general form of the equations of motion with respect to an exchange of the indices 1 and 2 belonging to $\mathcal{S}_1$ and $\mathcal{S}_2$ is assumed. On the other hand, an *asymmetry* between $\mathcal{S}_1$ and $\mathcal{S}_2$ will come about in two ways:

1. The trend parameters of $\mathcal{S}_1$ and $\mathcal{S}_2$ can have different values due to their different collective psychologies or economic and political structures.

2. Even for trend parameters that are numerically equal in both societies, the solutions of the equations need not be symmetrical with respect to both systems.

For our "coarse" model only three primary grossvariables for each system $\mathcal{S}_j$ are assumed:

1. the *variable of political psychology* P_j describing the state of the "political atmosphere" in $\mathcal{S}_j$ with respect to the other macrosociety and ranging from "sympathy" over "coexistence" to "tension",

2. the *equipment variable* E_j being a measure of the total disposable equipment in $\mathcal{S}_j$,

3. the *destructiveness variable* Q_j describing the degree of "readiness" of the society $\mathcal{S}_j$ to inflict destruction on the other society in the case of war. By definition $Q_j = 0$ in the case of peace. Functions of P_j, E_j and Q_j ($j = 1, 2$) will be introduced below using additional model variables.

The procedure of introducing the variable P_j is now described by giving a short account of the general ideological background in which P_j is embedded, and by setting up a method for quantifying the "state of political psychology with respect to the competing macrosystem".

Ideological Background. Every macrosociety is regulated by ideas on and valuations of the aims of individual life and the means of achieving them, and therefore of the ensuing social, economic and political structures. The guiding principles are a set of collective macrovariables playing the role of a multidimensional order parameter in the society. A distinction between "soft" and "hard" order parameters can be made.

Soft order parameters characterize flexible societies open to an exchange of ideas and offering a large range of intellectual and material options to their members. However, a society with order parameters that are too soft may degenerate into an orientationless or even chaotic state.

Hard order parameters characterize a rigid society reacting prohibitively against an exchange of ideas and prescribing a narrow range of intellectual and material options for its members. A society with order parameters that are too hard may degenerate into a self-isolating totalitarian system with a dogmatic ideology.

The interaction of macrosocieties may have either a profitable or an endangering mutual effect on their substance including their internal guiding principles. Hence, an interaction can elicit reactions with a strong emotional component.

Among the guiding ideological grossvariables of $\mathcal{S}_j$ the variable P_j is, by definition, the one characterizing the state of this interaction with the competing system. Therefore, the interaction dynamics to be described by the equations of motion will have to reflect the possible emotional background. The main effects of the latter are the properties of *self-acceleration* and *self-saturation*. This implies, for example, that tension generates tension and détente generates détente. On the other hand, tension as well as détente do not increase indefinitely but reach a saturation point. These properties of P_j will be further discussed *when setting up the equations of motion* in Sect. 6.2.2.

Quantification Method. The variable P_j can now be quantified: It is well known that in competitive societies certain relative attitudes develop towards the other society. These attitudes not only reflect political opinions in the narrow sense but also include attitudes valuating their own and the other system in all cultural, economic and ideological respects.

Ordering these attitudes according to their contribution to the political climate existing with respect to the competing system, a "dove/hawk" attitude model can be constructed. The attitude scale ranges from extreme to mildly *dovish,* namely from appeasing, pacifistic, compromising, conciliative, cooperative, moderative to understanding attitudes up to those moderately to extreme *hawkish*, namely self-respecting, patriotic, ideologically insistent, dogmatic, antagonistic, confrontative or even militant attitudes.

Here the attitudes are characterized by words with a deliberate emotional touch, as it is the emotional component of attitudes which determines the political atmosphere. This does not exclude that the attitudes so described also have sober and rational ingredients.

It is to be expected that hawkish attitudes give rise to "tension" by advocating definite, even militant standpoints and actions, while dovish attitudes give rise to "détente" by advocating conciliation, negotiation and compromise. $(2L + 1)$ characteristic attitudes $a = (-L), (-L+1), \ldots, (-1), 0, 1, \ldots, L$ are now selected, where increasingly positive (negative) indices $a > 0$ $(a < 0)$ denote dovish (hawkish) attitudes of increasing intensity and resoluteness. Furthermore it will be assumed that the selection of attitudes allows an allocation of one of the $(2L + 1)$ attitudes uniquely to each member of the society $\mathcal{S}_j$ at a given point in time. If then $n_a^{(j)}$ members of the society $\mathcal{S}_j$ have the attitude a, the following dove/hawk socio-configuration in $\mathcal{S}_j$ is obtained:

$$\{n_{-L}^{(j)}, n_{-L+1}^{(j)}, \ldots, n_{-1}^{(j)}, n_0^{(j)}, n_1^{(j)}, \ldots, n_{L-1}^{(j)}, n_L^{(j)}\} \tag{6.1}$$

where

$$\sum_{a=-L}^{+L} n_a^{(j)} = N_j$$

is the number of members of $\mathcal{S}_j$.

Making use of the socio-configuration (6.1) it is now possible to define *a collective variable P_j of political psychology in $\mathcal{S}_j$* ranging rom (-1) to $(+1)$ where increasingly positive (negative) values represent an increasingly dovish (hawkish) socio-configuration. In defining P_j dovish and hawkish attitudes are treated symmetrically. Since P_j is to be defined as an indicator of the intensity of détente (positive values) and of tension (negative values), and since it is reasonable to suppose that individuals contribute the more to détente (tension), the more resolute their dovish (hawkish) attitude is, increasing positive (negative) weights should be allocated to attitudes of increasing positive (negative) index. These requirements are satisfied, if in the definition of P_j, viz:

$$P_j = \sum_{a=-L}^{+L} g_a^{(j)} n_a^{(j)} \quad \text{with} \quad (-1) \leq P_j \leq 1 , \tag{6.2}$$

the weighing factors g_a allocated to attitudes a are chosen as follows:

$$\left.\begin{array}{ll} \text{a)}\ 0 < g_1^{(j)} < g_2^{(j)} < \ldots < g_L^{(j)} & \\ \text{b)}\ g_{-a}^{(j)} = -g_a^{(j)} ; & a = 0, 1, 2, \ldots, L \\ \text{c)}\ g_L^{(j)} = |g_{-L}^{(j)}| = \dfrac{1}{N_j} . & \end{array}\right\} \tag{6.3}$$

Clearly, there exist many possible definitions of P_j corresponding to different choices of the weighing factors $g_a^{(j)}$. However, this arbitrariness is not too relevant, as will be seen later, in that a whole class of grossvariables P_j obeys the same kind of equations of motion under appropriate assumptions for the transition probabilities.

The next problem is the definition of E_j. In contrast to the "abstract" and *intensive* variable P_j of *collective psychology*, the equipment variable E_j is a *collective material* variable. This variable is *extensive*, and therefore proportional to the size of the system. In seeking a plausible definition of E_j one can start from the gross national product Y_j, i.e. the total value of goods and services produced in $\mathcal{S}_j$ per year, which is conventionally decomposed into

$$\mathrm{Y}_j = \mathrm{C}_j + \mathrm{J}_j + \mathrm{S}_j \tag{6.4}$$

where C_j denotes the value of all consumer goods per year, J_j all investments per year and S_j all state expenditures including services, armament, infrastructure facilities, etc. produced per year. It τ_C, τ_J and τ_S are the mean life times (in years) of the components C_j, J_j and S_j the total equipment E_j can be defined by

$$E_j = \mathrm{C}_j \tau_\mathrm{C} + \mathrm{J}_j \tau_\mathrm{J} + \mathrm{S}_j \tau_\mathrm{S} \ . \tag{6.5}$$

It is irrelevant in which currency E_j is measured since appropriately scaled dimensionless variables will be used in the equations of motion.

In the context of competitive macrosocieties an *alternative decomposition* of E_j into *"wealth"* W_j (= value of all civil goods and services including the investment necessary to produce them) and *"armament"* A_j (= value of total military equipment and services including the corresponding necessary investment), i.e.,

$$E_j = W_j + A_j \tag{6.6}$$

[where $W_j = w_j E_j$; $A_j = a_j E_j$ and $(a_j + w_j) = 1$] seems to be more relevant for the collective material view of the political climate existing between $\mathcal{S}_1$ and $\mathcal{S}_2$. In the minimal model being discussed here the coefficients w_j and a_j of civil and military equipment, respectively, will be treated as *given parameters*. It is possible, however, to extend the model by treating W_j and A_j as *independent dynamic variables* and, for instance, by explicitly considering the progressive relationship between W_j and A_j in interaction with P_j $(j = 1, 2)$.

Lastly, the relation between the collective psychology variable *"destructiveness"* Q_j of $\mathcal{S}_j$ and the material variable *"destruction rate"* D_i (= equipment destroyed per time unit in $\mathcal{S}_i$) in the case of war has to be established. On a "coarse-grained" level of argumentation this relation can be defined as follows: The destruction rate D_2 inflicted on society $\mathcal{S}_2$ in absolute terms is (roughly) proportional to:

1) the destructiveness Q_1 (i.e. the readiness of $\mathcal{S}_1$ to apply its military power against $\mathcal{S}_2$),

2) the "armament" $A_1 = a_1 E_1$ which is the total military equipment or power of $\mathcal{S}_1$,

3) the amount of collective material E_2 in $\mathcal{S}_2$ or parts of it, like $W_2 = w_2 E_2$ and/or $A_2 = a_2 E_2$, which could be destroyed due to military action by $\mathcal{S}_1$.

The following relationships appear, therefore, to be appropriate

$$D_1 = \varrho_1 E_1 E_2 Q_2 \tag{6.7a}$$

$$D_2 = \varrho_2 E_1 E_2 Q_1 \,, \tag{6.7b}$$

where ϱ_1, ϱ_2 are open coefficients depending on the situation in which warfare occurs.

6.2.2 The Equations of Motion

The equations of motion for the system described in Sect. 6.2.1 cannot be derived from fundamental microscopic theories and only partially from direct measurements. Also too simple-minded analogies to physical systems and models fail to provide good guiding principles for a derivation. One can only rely on political and economical insight and on acquired knowledge in translating this in terms of coupled non-linear differential equations for the grossvariables. Term by term this translation has to be explained and justified on the basis of qualitative reasoning on the interaction of economic and political influences. Hence, model equations such as those here formulated are certainly open to criticism. A valuable criticism could, for instance, consist of proposing modified or additional terms in the equations of motion or in formulating further equations involving additional variables. If the original model is "stable" and "good" with respect to the modelled interactions, a modified or extended model will lead to a somewhat shifted or refined picture without fundamentally changing the global features of the original conception[3]. In any case, any form of criticism would help to establish the framework of a "feedback between qualitative and quantitative argument" as described in Sect. 6.1.

On the other hand, this book would not have been written – and particularly this chapter – if the authors were essentially to agree with a global criticism claiming that "matters as complex as the policy of macrosocieties in principle cannot be represented in terms of a mathematical model, i.e. by a set of equations of motion". On the contrary it is the authors impression that even the *most* elaborate qualitative analyses deal at most with a few dozen eco-political macrovariables, their interrelations and time dependence, using another few dozen trend estimates of the probable behaviour of populations and their respective leaders. It is just this kind of analysis which – in principle – can be given a quantitative descriptive form.

Turning to the equations of motion for the six primary gross variables P_1, P_2, E_1, E_2, Q_1 and Q_2 it is convenient to begin with the variables P_j of political psychology in $\mathscr{S}_j$. In the simplest non-trivial case the socio-configuration (6.1) would consist of members of the society disposing of *one* dovish ($a = 1$) and *one* hawkish ($a = -1$) attitude only:

$$\{n_1^{(j)}; n_{-1}^{(j)}\} \qquad \text{with} \quad n_1^{(j)} + n_{-1}^{(j)} = N_j \,. \tag{6.8}$$

[3] For an extensive discussion...[6.8]

The definition (6.2) then leads to

$$\left.\begin{aligned} P_j &= \frac{n_1^{(j)} - n_{-1}^{(j)}}{N_j}\,; \qquad -1 \leq P_j \leq +1 \\ n_1^{(j)} &= \frac{N_j}{2}(1 + P_j); \qquad n_{-1}^{(j)} = \frac{N_j}{2}(1 - P_j)\,. \end{aligned}\right\} \tag{6.9}$$

Assuming individual transition probabilities between the two attitudes of the form

$$p_{1,-1}^{(j)} = \nu \exp(u_j)\,; \qquad p_{-1,1}^{(j)} = \nu \exp(-u_j) \tag{6.10}$$

where the u_j can be functions of all the grossvariables P_i, E_i, Q_i, the following mean value equations can be derived

$$\left.\begin{aligned} \frac{dn_1^{(j)}}{dt} &= -\,p_{-1,1}^{(j)}\, n_1^{(j)} + p_{1,-1}^{(j)}\, n_{-1}^{(j)} \\ \frac{dn_{-1}^{(j)}}{dt} &= p_{-1,1}^{(j)}\, n_1^{(j)} - p_{1,-1}^{(j)}\, n_{-1}^{(j)} \end{aligned}\right\} \tag{6.11}$$

or equivalently, using (6.9, 10), as

$$\frac{dP_j}{dt} = 2\nu(\sinh u_j - P_j \cosh u_j)\,. \tag{6.12}$$

This equation corresponds to the mean value equation (2.37 c) of the opinion formation model in Chap. 2 except for the dependence of u_j on further gross variables.

Of course, the assumption of *one* dovish and *one* hawkish attitude only is an oversimplification. Returning to the general case described by (6.1–3) it can be shown that there exists a plausible choice of individual transition probabilities between the elements of a set of gradually varying attitudes leading to *exactly the same equation of motion* (6.12) for the collective variable P_j, even though the socio-configuration considered is much more complicated.

This plausible choice for the transition probabilities is given by (leaving out the index $j = 1, 2$ for the moment)

$$\left.\begin{aligned} p_{(a+1)\leftarrow a} &\equiv p_{a\uparrow} = f_{a\uparrow} \cdot \exp(u) \\ p_{(a-1)\leftarrow a} &\equiv p_{a\downarrow} = f_{a\downarrow} \cdot \exp(-u) \\ p_{a'\leftarrow a} &\quad\;\, = 0 \qquad \text{for} \quad a' \neq a \pm 1 \end{aligned}\right\} \tag{6.13}$$

which is to be interpreted as follows:

1) Only transitions from one attitude to the neighbouring attitude in the sense of the ordered scheme are allowed.

2) The dependence of the upward and downward transition probabilities on the socio-configuration is expressed by the factors $\exp(u)$ and $\exp(-u)$ only, where u is a function of the *collective variables* P, E and Q. This assumption can be justified in that modern media can be thought of as representing the collective states described by P, E and Q and simultaneously as acting as a form of order parameter strongly influencing individual opinion and the probabilities of a change of opinion.

3) The pre-multiplying parameters $f_{a\uparrow}$ and $f_{a\downarrow}$ are constant "strength" factors characteristic for the upward and downward transitions from attitude a.

Inserting (6.13) into the mean value equations of motion for the components n_a of the socio-configuration,

$$\frac{dn_a}{dt} = p_{(a-1)\uparrow} n_{(a-1)} + p_{(a+1)\downarrow} n_{(a+1)} - (p_{a\uparrow} + p_{a\downarrow}) n_a$$

$$= s_a(n)\sinh(u) - c_a(n)\cosh(u) \tag{6.14}$$

is obtained, where

$$\left.\begin{aligned} s_a(n) &= -[(f_{a\uparrow} n_a - f_{(a-1)\uparrow} n_{(a-1)}) + (f_{(a+1)\downarrow} n_{(a+1)} - f_{a\downarrow} n_a)] \\ c_a(n) &= [(f_{a\uparrow} n_a - f_{(a-1)\uparrow} n_{(a-1)}) - (f_{(a+1)\downarrow} n_{(a+1)} - f_{a\downarrow} n_a)]\ . \end{aligned}\right\} \tag{6.15}$$

From (6.2) and (6.14) the differential equation for the grossvariable P of political psychology

$$\frac{dP}{dt} = \sum_{a=-L}^{+L} g_a \frac{dn_a}{dt} = S\sinh(u) - C\cosh(u) \tag{6.16}$$

follows, where

$$S = \sum_{a=-L}^{+L} g_a s_a(n) \tag{6.17}$$

$$C = \sum_{a=-L}^{+L} g_a c_a(n)\ . \tag{6.18}$$

In general, the expression on the rhs of (6.16) not only depends on the collective variables P, E, and Q but, due to the presence of the factors S and C, also on the whole socio-configuration. To obtain the equivalence of (6.16) with the simple dove/hawk model equation (6.12), choices for the open parameters $f_{a\uparrow}$ and $f_{a\downarrow}$ have to be made, which yield

$$S = 2\nu = \text{const} \quad \text{and} \quad C = 2\nu P\ . \tag{6.19}$$

In order to decide whether these choices exist, a quasi-continuous sequence of attitudes can be assumed and the summations (6.17, 18) evaluated as integrals.

Introducing

$$f_+(a) = (f_{a\uparrow} + f_{a\downarrow}) \qquad \text{and} \quad f_-(a) = (f_{a\downarrow} - f_{a\uparrow}) \tag{6.20}$$

and using (6.15) as continuous forms, instead of (6.17, 18)

$$S \approx \int_{-L}^{+L} g(a)\, s_a(n)\, da$$

$$= -\int_{-L}^{L} g(a) \frac{\partial}{\partial a} [f_+(a)\, n(a)]\, da = +\int_{-L}^{L} g'(a)\, [f_+(a)\, n(a)]\, da \tag{6.21}$$

and

$$C \approx \int_{-L}^{L} g(a)\, c_a(n)\, da$$

$$= -\int_{-L}^{L} g(a) \frac{\partial}{\partial a} [f_-(a)\, n(a)]\, da = +\int_{-L}^{L} g'(a)\, [f_-(a)\, n(a)]\, da \tag{6.22}$$

with $g'(a) \equiv \partial g(a)/\partial a$, are obtained. In the partial integrations leading to the last expressions in (6.21, 22), it can be assumed that the boundary terms may be neglected, as the number of individuals $n(L)$, $n(-L)$ with the extreme attitudes $L, (-L)$ is small. Now it can be seen that (6.19) and hence (6.12) can be realised by the choice of

$$f_-(a) = \frac{g(a)}{g'(a)}\, 2\nu \qquad \text{and} \quad f_+(a) = \frac{2\nu}{N g'(a)} \tag{6.23}$$

or

$$f_\downarrow(a) = \frac{\nu}{N g'(a)}\, [1 + N g(a)] \qquad \text{and}$$

$$f_\uparrow(a) = \frac{\nu}{N g'(a)}\, [1 - N g(a)]$$

where $f_\downarrow(a)$ and $f_\uparrow(a)$ – because of (6.3 c) – automatically satisfy the boundary conditions

$$f_\downarrow(-L) = 0 \qquad \text{and} \quad f_\uparrow(+L) = 0\,. \tag{6.24}$$

For example, defining P according to (6.2) with equidistant spacing of the weights

$$g(a) = \frac{a}{NL} \quad \text{with} \quad -L \leq a \leq +L \tag{6.25}$$

it can be seen from (6.23) that the choice of the parameters

$$f_{\uparrow}(a) = \nu(L - a) \quad \text{and} \quad f_{\downarrow}(a) = \nu(L + a) \tag{6.26}$$

in the transition probabilities (6.13) leads to the equation of motion (6.12) for P_j.

The choice (6.26) is plausible in itself. It means that, as far as the parameters f are concerned, the transition to higher levels of the attitude a is less probable than the transition to lower levels if one starts from a positive attitude $a > 0$, and vice versa for attitudes starting with $a < 0$. The above considerations show that the reasonable assumptions of (6.13) for the transition probabilities together with the choice (6.23) lead to the simple equation of motion (6.12) for the collective variable P, as defined by (6.2), even if the underlying socio-configurations have a complex structure with a sequence of many intermediate attitudes between dovishness and hawkishness. It will be assumed, therefore, that the equations of motion for $P_j(t)$ are of the form (6.12).

In order to make this equation fully explicit, the form of u_j as a function of P_1, P_2, E_1, E_2, Q_1 and Q_2 has to be specified.

It is obvious, that in particular the self-accelerating and self-saturating properties of P_j mentioned above have to be included in the dependence of u_j on P_j. It will now be seen that the simple form for u_j with respect to its dependence on P_j, namely

$$\begin{aligned} u_1(P_1, P_2, E_1, E_2, Q_1, Q_2) &= \delta_1 + \kappa_1 P_1 \\ u_2(P_1, P_2, E_1, E_2, Q_1, Q_2) &= \delta_2 + \kappa_2 P_2 \end{aligned} \tag{6.27}$$

already takes account of these self-influence effects – the dependence of δ_j and κ_j on the remaining variables will be specified later.

Introducing "potentials" $V_1(P_1, \ldots)$ and $V_2(P_2, \ldots)$ by writing (6.12) in the form

$$\frac{dP_1}{dt} = 2\nu[\sinh(u_1) - P_1 \cosh(u_1)] = -\frac{\partial V_1}{\partial P_1} \tag{6.28 a}$$

$$\frac{dP_2}{dt} = 2\nu[\sinh(u_2) - P_2 \cosh(u_2)] = -\frac{\partial V_2}{\partial P_2} \tag{6.28 b}$$

the potentials

$$V_1(P_1, \ldots) = -\frac{2\nu}{\kappa_1^2}[(1+\kappa_1)\cosh(\delta_1+\kappa_1 P_1) - \kappa_1 P_1 \sinh(\delta_1+\kappa_1 P_1)] + c_1 \tag{6.29 a}$$

$$V_2(P_2, \ldots) = -\frac{2\nu}{\kappa_2^2}[(1+\kappa_2)\cosh(\delta_2+\kappa_2 P_2) - \kappa_2 P_2 \sinh(\delta_2+\kappa_2 P_2)] + c_2 \tag{6.29 b}$$

are obtained. In Fig. 6.1 the potential $V_j(P_j)$ is plotted as a function of P_j for a given value $\kappa_j > 1$ and for different values of δ_j.

Equations (6.28 a, b) then describe the motion of P_j with this potential under consideration of the restoring force $(-\partial V_j/\partial P_j)$. Considering for the moment κ_j and δ_j as constants, this motion comes to rest at the potential minima. For large positive $\delta_j \geq \delta_{cj}$ [see definition (6.31)] – namely, with a high preference for détente – there exists only one minimum of $V_j(P_j)$ at $P_{j+} > 0$ describing an unambiguous stationary state of détente in which the system $\mathcal{S}_j$ is stable. For smaller but still positive $\delta_j < \delta_{cj}$ – indicating a low preference for détente – there exist two minima for $V_j(P_j, \ldots)$, a relatively deep minimum at $P_{j+} > 0$ and a shallow one at $P_{j-} < 0$; for $\delta_j = 0$ the potential $V_j(P_j, \ldots)$ becomes symmetrical and both minima have the same depth. The case $0 < \delta_j < \delta_{cj}$ describes an ambiguity of the possible developments: Depending on the initial conditions, the variable P_j may either develop to a state $P_{j+} > 0$ of détente or to a state $P_{j-} < 0$ of tension due to the effects induced by P_j on itself. Because of the symmetry of the potentials

$$V_j(|P_j|; |\delta_j|, \kappa_j) = V_j(-|P_j|; -|\delta_j|, \kappa_j) \tag{6.30}$$

the analogous situations with $\delta_j < 0$ (namely, with a preference for tension) arise, i.e., an ambiguity with two possible stationary states at $P_{j-} < 0$ and $P_{j+} > 0$ for $|\delta_j| < |\delta_{cj}|$ and an unambiguous state of tension $P_{j-} < 0$ for $|\delta_j| > |\delta_{cj}|$. The critical value of δ_{cj} separating these cases is given by

$$\cosh^2(|\delta_{cj}| - \sqrt{\kappa_j(\kappa_j - 1)}) = \kappa_j\,, \qquad \text{where} \quad \kappa_j > 1\,. \tag{6.31}$$

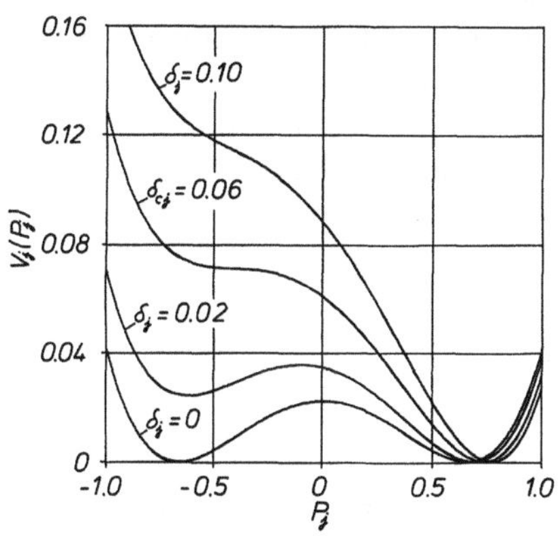

Fig. 6.1. The potential $V_j(P_j)$ for the value $\kappa_j = 1.2$ and for $\delta_j = 0$, 0.02, 0.06 and 0.10. The transition from a bistable to a monostable potential shape takes place at $\delta_{cj} = 0.06$

So far, the discussion of the equation of motion (6.28) and of the potentials formally corresponds with that of the equations (2.48) and (2.124) for the opinion formation model in Sect. 2.4. In the presently discussed model, however, P_j is a collective variable of a complex socio-configuration and, more important, the trend parameters $\{\delta_j, \kappa_j\}$ depend on further dynamic grossvariables. Hence the shape of the potentials (6.29) will vary during the motion of all grossvariables. The equations (6.28) then describe the motion of P_j in terms of time dependent potentials, whose shape is determined by the other equations of motion not yet discussed. Thus the form of the dependence of the trend parameters $\{\delta_j, \kappa_j\}$ on the other grossvariables should now be considered.

The magnitude of the parameter $\kappa_j > 0$ describes the strength of the "self-influence" of P_j on decisions about dove/hawk attitudes in the society $\mathcal{S}_j$. As a first approximation κ_j can be considered as an *intrinsic* property of $\mathcal{S}_j$ arising from the average "mentality" of its members. It can therefore be assumed that κ_j is *independent* of the grossvariables $\{P_1, P_2, E_1, E_2, Q_1, Q_2\}$ in the model. It is certainly a plausible assumption that an open society with members who are largely independent and individualistic in their thinking has a small κ, while a self-isolating society with a high internal propaganda level and collectivistic political institutions has a large κ.

The parameter $\delta_j \gtrless 0$ describes the strength or magnitude and the sign of a general decision preference in $\mathcal{S}_j$ for détente ($\delta_j > 0$) or tension ($\delta_j < 0$) with respect to the competing society. The value of δ_j will depend not only on the intrinsic "mentality" within $\mathcal{S}_j$ but also on a range of ideological and economic influences. The dependence of δ_j on the material and psychological grossvariables introduced above will now be considered, however, an extension of the present model including further internal or external influences on δ_j certainly seems to be feasible.

The general form assumed for δ_j is

$$\left.\begin{aligned} \delta_1 &= \delta_{10} + \delta_{1E} + \delta_{1P} + \delta_{1D} \\ \delta_2 &= \delta_{20} + \delta_{2E} + \delta_{2P} + \delta_{2D} \end{aligned}\right\} \qquad (6.32)$$

where the terms δ_{j0}, δ_{jE}, δ_{jP}, and δ_{jD} describe the respective contributions to δ_j of the intrinsic détente/tension preferences, of the economy, of the variable P_i of political psychology of the competing society $\mathcal{S}_i$ and (in the case of war) of the destruction rate D_j inflicted on $\mathcal{S}_j$ by the hostile society $\mathcal{S}_i$.

An immediate task is to find expressions for δ_{jE}, δ_{jP} and δ_{jD} which are plausible and which are also flexible enough to admit of a variety of possible interactions.

The expressions

$$\left.\begin{aligned} \delta_{1E} &= \alpha_{11} E_1 - \alpha_{12} E_2 \\ \delta_{2E} &= \alpha_{22} E_2 - \alpha_{21} E_1 \end{aligned}\right\} \qquad (6.33)$$

allow a simple description of the influence of the economy of both systems on the dove/hawk preferences in the two societies $\mathcal{S}_j$. The choice of the parameters α_{jj}, $\alpha_{ji}\,(i \neq j)$ then requires a philosophy of its own. For instance: Choosing positive α_{jj} would imply that "increasing economic standards in $\mathcal{S}_j$ promotes dovish attitudes in $\mathcal{S}_j$" and choosing positive $\alpha_{ji}\,(i \neq j)$ would then mean that "increasing economic standards in the competing system $\mathcal{S}_i$ tends to further hawkish attitudes in $\mathcal{S}_j$". In a more refined theory a distinction could be made between the influence of wealth W and armaments A in and on $\mathcal{S}_1$ and $\mathcal{S}_2$, as introduced in (6.6), on the preference parameter δ_j.

The forms

$$\left.\begin{aligned} \delta_{1P} &= \kappa_{12} P_2 \\ \delta_{2P} &= \kappa_{21} P_1 \end{aligned}\right\} \tag{6.34}$$

are the simplest which take into account the influence of the collective political psychology of one society on the dove/hawk preferences of the other society. The terms (6.34) and the self-influence terms $\kappa_j P_j$ in u_j, see (6.27), comprise the same type of influence, namely the effect of internal and external political psychology variables on their own dynamics. Again the choice of the parameters $\kappa_{ij}\,(i \neq j)$ requires a philosophy of its own:

If, for instance, the assumption proves to be correct that the political psychology of an open society, say $\mathcal{S}_1$, reacts sensitively on the political state of the competing system $\mathcal{S}_2$, a large value for κ_{12} can be expected. Even the sign of κ_{12} can be appropriately chosen; a positive (negative) κ_{12} would mean that hawkishness in $\mathcal{S}_2$ induces a hawkish (dovish) trend in $\mathcal{S}_1$. If, on the other hand, the assumption is valid that a self-isolating totalitarian system, say $\mathcal{S}_2$, wants to protect its internal ideology from external influences coming from $\mathcal{S}_1$, a small value of κ_{21} or a value strongly dominated by internal propaganda within $\mathcal{S}_2$ has to be expected.

In the case of warfare the additional terms δ_{jD} appear in the expression for δ_j. For this contribution the following is proposed:

$$\left.\begin{aligned} \delta_{1D} &= -(\beta_{11} D_1 + \beta_{12} Q_1) + \left(\frac{E_{1S}}{E_1}\right) \\ \delta_{2D} &= -(\beta_{22} D_2 + \beta_{21} Q_2) + \left(\frac{E_{2S}}{E_2}\right). \end{aligned}\right\} \tag{6.35}$$

The first terms in δ_{jD}, implying a strong preference for "tension" and hawkish attitudes in $\mathcal{S}_j$ against $\mathcal{S}_i$, are due firstly to the destruction rate D_j inflicted – through any kind of warfare conducted by the other society $\mathcal{S}_i$ – and, secondly, to the enhancement effect of the ensuing war readiness Q_j in society $\mathcal{S}_j$. Since D_j and Q_j are positive semi-definite by definition, such an effect is described when β_{jj} and β_{ji} are chosen as positive coefficients. On the other hand, the second terms in δ_{jD} are meant to describe the effect of exhaustion and war weariness in $\mathcal{S}_j$ leading to a breakdown of hawkish attitudes and a definitive reappearance of conciliatory preferences when the total equipment level E_j in $\mathcal{S}_j$ sinks (because of destruction) below a certain "survival threshold value", E_{js}.

The equations (6.28 a, b) with u_j given by (6.27) and δ_j specified by (6.32–35) complete the description of the dynamics of P_j.

Turning now to the dynamics of the equipment variable E_j, it should be observed that since E_j is a global material variable only so its equation of motion is not as difficult to construct as that for P_j. It has the form of a rate of change equation, where the terms on the rhs describe positive or negative contributions to the change of E_j in each time unit. After transition to the dimensionless scaled time variable

$$\tau = 2\nu t \tag{6.36}$$

the following forms are assumed

$$\varepsilon \frac{dE_1}{d\tau} = (E_1 - E_{10}^{-1} E_1^2) + \nu_1 P_1 E_1 + (\gamma_{11} P_1 + \gamma_{12} P_2) E_1 E_2 - D_1 \tag{6.37 a}$$

and

$$\varepsilon \frac{dE_2}{d\tau} = (E_2 - E_{20}^{-1} E_2^2) + \nu_2 P_2 E_2 + (\gamma_{21} P_1 + \gamma_{22} P_2) E_1 E_2 - D_2 \,. \tag{6.37 b}$$

The destruction rates D_1, D_2 per unit of time on the rhs of (6.37 a, b) then refer to the time variable $\tilde{\tau} = \tau/\varepsilon$. Disposing finally of the (as yet open) scale of the "destructiveness variables" Q_1, Q_2 the coefficients ϱ_1, ϱ_2 in (6.7 a, b) may be put equal to unity, so that the final relations between D_1, D_2 and Q_2, Q_1 read

$$D_1 = E_1 E_2 Q_2 \tag{6.38 a}$$

$$D_2 = E_1 E_2 Q_1 \,. \tag{6.38 b}$$

An interpretation of the terms of (6.37) can be made along the following lines:

The scaling factor ε takes account of the time scale upon which changes in the material variable E_j are registered as taking place. It can be expected that material variables exhibit change typically at a slower rate than the variables of political psychology, whose rate of change ν is used in defining the dimensionless time variable τ – one possible sad exeption here may be the high destruction rates to be associated with ultramodern warfare.

The first term $(E_j - E_{j0}^{-1} E_j^2)$ of the rhs describes the *expansion* of the total equipment as being proportional to its own capacity but with a *saturation* term. It is plausible to assume the saturation term as being proportional to E_j^2 because it arises from the "internal friction" between the contributions to the growing equipment level caused by their mutual limitations. Without any further influences this first term would describe the development of the E_j towards the stationary level E_{j0}.

The second term $\nu_j P_j E_j$ and the third term $(\gamma_{jj} P_j + \gamma_{ji} P_i) E_i E_j$ postulate a modification of the growth and saturation process due to the influence of the psychology variables and the interaction between the economies of the two societies. These terms, being proportional to E_j, lead to a constructive or destructive and in general time dependent renormalization of the growth rate and the saturation level of E_j. The choice of positive values for the parameters ν_j, γ_{jj} and γ_{ji} seems to be quite plausible.

The last term in (6.37) vanishes in normal "peaceful" times being, by definition, the destruction rate, i.e. the amount of destroyed equipment per unit time, measured in $\bar{\tau}$ in the case of warfare.

In a more refined model *separate* equations of motion for variables "wealth" W_j and "armament" A_j could be established. Such an extended model could – in contrast to the present version – take into account mutual reactions in $\mathcal{S}_1$ and $\mathcal{S}_2$ to shifts in the ratio of civil to military production.

To complete the model the equation of motion for the destructiveness variable Q_j has to be established. Although being a variable of collective psychology in $\mathcal{S}_j$ it is directly connected with the material destruction rate D_i by (6.38). The equations for Q_1 and Q_2 should reflect the following principal characteristics of collective psychology in the eventuality of war occuring between the two societies:

a) Any destruction rate D_j inflicted on $\mathcal{S}_j$ by the society $\mathcal{S}_i$ will provoke an increase in the destructiveness variable Q_j in the society $\mathcal{S}_j$, and vice versa, hence opening a process of mutual escalation of violence.

b) Competing with this effect, under normal "reason-dominated" circumstances there will exist a strong damping influence of the internal collective psychology P_j in $\mathcal{S}_j$ on the destructiveness variable Q_j anticipating the negative results of an ensuing escalation. Only, if P_j crosses a strongly hawkish threshold value $P_{j\text{th}} < 0$ in the direction of extremely hawkish values, the damping effect of P_j is reversed into an reinforcement effect on Q_j.

Both characteristics are taken into account in the following equations:

$$\begin{aligned}\frac{dQ_1}{d\tau} &= \mu_{11}(P_{1\text{th}} - P_1)\,Q_1 + \mu_{12} D_1\\ &= a_{11} Q_1 + a_{12} Q_2 \qquad \text{with} \qquad (6.39\,\text{a})\\ &\quad a_{11} = \mu_{11}(P_{1\text{th}} - P_1) \qquad \text{and} \quad a_{12} = \mu_{12} E_1 E_2\end{aligned}$$

and

$$\begin{aligned}\frac{dQ_2}{d\tau} &= \mu_{22}(P_{2\text{th}} - P_2)\,Q_2 + \mu_{21} D_2\\ &= a_{21} Q_1 + a_{22} Q_2 \qquad \text{with} \qquad (6.39\,\text{b})\\ &\quad a_{22} = \mu_{22}(P_{2\text{th}} - P_2) \qquad \text{and} \quad a_{21} = \mu_{21} E_1 E_2\end{aligned}$$

where μ_{11}, μ_{12}, μ_{21} and μ_{22} are positive coefficients. Obviously, (6.39 a, b) are linear homogeneous equations in Q_j with, in general, time dependent coefficients and $Q_1 = Q_2 = 0$ as a trivial (stable or unstable) solution. To decide whether the rate of change of Q_j is escalating or de-escalating one may treat a_{ij} as constants at the given moment in time and solve (6.39) under this assumption as

$$Q_j(t) = c_+ Q_{j+} \exp(\lambda_+ t) + c_- Q_{j-} \exp(\lambda_- t) \tag{6.40}$$

with eigenvalues

$$\lambda_\pm = 1/2 \left[(a_{11} + a_{22}) \pm \sqrt{(a_{11} - a_{22})^2 + 4 a_{12} a_{21}}\right]. \tag{6.41}$$

Because $a_{12} > 0$ and $a_{21} > 0$, the eigenvalues are always real. A de-escalating situation is only realized for

$$\lambda_+ < 0 \quad \text{and} \quad \lambda_- < 0 \tag{6.42}$$

or, equivalently, for

$$a_{11} < 0, a_{22} < 0 \quad \text{and} \quad a_{11} a_{22} > a_{12} a_{21}, \tag{6.43 a}$$

that is, for

$$P_1 > P_{1\text{th}} \quad \text{and} \quad P_2 > P_{2\text{th}} \tag{6.43 b}$$

together with $\mu_{11}\mu_{22}(P_{1\text{th}} - P_1)(P_{2\text{th}} - P_2) > \mu_{12}\mu_{21} E_1^2 E_2^2$. It must, of course, be kept in mind that it depends on the change of $P_j(t)$ and $E_j(t)$ with time via the coupled equations of motion, whether and when the conditions (6.43) are fulfilled or not. Hence only the total situation decides whether given initial non-vanishing destructiveness variables and the ensuing destruction rates develop into escalating or de-escalating states.

Summarizing the model:

1) The equations for the variables of political psychology P_j (6.28) are

$$\frac{dP_j}{d\tau} = \sinh(u_j) - P_j \cosh(u_j)$$

where the scaled time variable is

$$\tau = 2\nu t$$

and where

$$u_j = \delta_{j0} + (\alpha_{jj} E_j - \alpha_{ji} E_i) + (\kappa_j P_j + \kappa_{ji} P_i)$$

$$- (\beta_{jj} D_j + \beta_{ji} Q_i) + \frac{E_{js}}{E_i}$$

2) The equations for the equipment variables E_j (6.37) are

$$\varepsilon \frac{dE_j}{d\tau} = (E_j - E_{j0}^{-1} E_j^2)$$

$$+ \nu_j P_j E_j + (\gamma_{jj} P_j + \gamma_{ji} P_i) E_i E_j - D_j \ .$$

3) The equations for the destructiveness variables Q_j (6.39) are

$$\frac{dQ_j}{d\tau} = \mu_{jj} (P_{j\mathrm{th}} - P_j) Q_j + \mu_{ji} D_j$$

where

$$D_j = E_i E_j Q_i \ .$$

In all equations $i, j = 1, 2$ and $i \neq j$.

6.3 Solutions of the Model Equations

The description of competitive macrosocieties given in Sect. 6.2 is certainly a "minimal" one. Nevertheless the model is of considerable complexity. It contains 30 trend parameters, 15 for each society, and two time scaling factors ν and ε, where ν is absorbed in the definition (6.36) of the scaled time. The trend parameters in (6.28, 37, 39), which have to be specified for a complete model description of the interaction of the societies $\mathcal{S}_1$ and $\mathcal{S}_2$, are as follows:

$$\left.\begin{array}{lll} \text{a)} & \delta_{j0},\ \alpha_{jj},\ \alpha_{ji},\ \kappa_j,\ \kappa_{ji},\ \beta_{jj},\ \beta_{ji} \text{ and } E_{js} & \text{belonging to (6.28),} \\ \text{b)} & \varepsilon,\ E_{j0},\ \nu_j,\ \gamma_{jj} \text{ and } \gamma_{ji} & \text{belonging to (6.37),} \\ \text{c)} & P_{j\mathrm{th}},\ \mu_{jj} \text{ and } \mu_{ji} & \text{belonging to (6.39).} \end{array}\right\} \quad (6.44)$$

The indices j and i are 1 or 2, where $i \neq j$. Because of this complexity a full account of the manifold of model solutions for all reasonable parameter combinations cannot be given here. However, in Sect. 6.3.1 the general structure of stationary solutions will be considered and in Sect. 6.3.2 a selection of numerical non-stationary solutions for a set of natural and interesting parameter combinations will be discussed.

6.3.1 Stationary Solutions

Setting the time derivatives $dP_j/d\tau$, $dE_j/d\tau$ and $dQ_j/d\tau$ in (6.28, 37, 39) to zero the six equations to be satisfied by stationary solutions $\overline{P}_j$, $\overline{E}_j$, $\overline{Q}_j$ of the model are obtained:

$$0 = \sinh(u_j[\overline{P}_l, \overline{E}_l, \overline{Q}_l]) - \overline{P}_j \cosh(u_j[\overline{P}_l, \overline{E}_l, \overline{Q}_l]) \quad (6.45)$$

$$0 = \overline{E}_j[(1 - E_{j0}^{-1}\overline{E}_j) + \nu_j \overline{P}_j + (\gamma_{jj}\overline{P}_j + \gamma_{ji}\overline{P}_i)\overline{E}_i - \overline{E}_i\overline{Q}_i] \quad (6.46)$$

$$0 = \mu_{jj}(P_{j\mathrm{th}} - \overline{P}_j)\overline{Q}_j + \mu_{ji}\overline{E}_j\overline{E}_i\overline{Q}_i \quad (6.47)$$

where $j = 1, 2$ and $i = 1, 2$ but $i \neq j$ in each equation. The two linear equations (6.47) for $\overline{Q}_1$ and $\overline{Q}_2$ would allow of "non-trivial" solutions $\overline{Q}_j \neq 0$ only when the determinant of the coefficients vanishes. Leaving this exceptional case aside a general non-vanishing determinant

$$[\mu_{11}\mu_{21}(P_{1\mathrm{th}} - \overline{P}_1) - \mu_{22}\mu_{12}(P_{2\mathrm{th}} - \overline{P}_2)]\overline{E}_1\overline{E}_2 \neq 0 \quad (6.48)$$

will be assumed leading to the only "trivial" stationary solution

$$\overline{Q}_1 = \overline{Q}_2 = 0 \; . \quad (6.49)$$

In other words, only "peace" can be stationary, while non-vanishing destructiveness Q_j and the ensuing destruction rates D_i (i.e. states of warfare) are principally non-stationary transient states.

On the other hand, reasonable stationary equipment values $\overline{E}_1$ and $\overline{E}_2$ have to be non-vanishing and positive definite by definition. Presuming (6.49) and $\overline{E}_j > 0$, the solution of (6.46) leads to

$$\overline{E}_1 = E_{10}\frac{(1 + \nu_1\overline{P}_1) + E_{20}(1 + \nu_2\overline{P}_2)(\gamma_{11}\overline{P}_1 + \gamma_{12}\overline{P}_2)}{1 - E_{10}E_{20}(\gamma_{11}\overline{P}_1 + \gamma_{12}\overline{P}_2)(\gamma_{21}\overline{P}_1 + \gamma_{22}\overline{P}_2)} \quad (6.50\,\mathrm{a})$$

$$\overline{E}_2 = E_{20}\frac{(1 + \nu_2\overline{P}_2) + E_{10}(1 + \nu_1\overline{P}_1)(\gamma_{21}\overline{P}_1 + \gamma_{22}\overline{P}_2)}{1 - E_{10}E_{20}(\gamma_{11}\overline{P}_1 + \gamma_{12}\overline{P}_2)(\gamma_{21}\overline{P}_1 + \gamma_{22}\overline{P}_2)} \quad (6.50\,\mathrm{b})$$

where, of course, (6.50 b) follows from (6.50 a) by exchange of the indices 1 and 2. The equations (6.50 a, b) together with the two transcendental equations (6.45) determine the stationary values for $\overline{E}_1$, $\overline{E}_2$, $\overline{P}_1$, and $\overline{P}_2$. The number of different solutions depends on the choice of the trend parameter sets a) and b) and will not be discussed here. But certain conditions for the trend parameters E_{j0}, ν_j, γ_{jj} and γ_{ji} can be derived from the range of the variables $\overline{P}_j$ and $\overline{E}_j$, namely

$$-1 \leq \overline{P}_j \leq +1 \quad \text{and} \quad 0 < \overline{E}_j < \infty \quad \text{where} \quad j = 1, 2 \; . \quad (6.51)$$

Inserting the extreme cases $\overline{P}_j = +1$ and $\overline{P}_j = -1$ [which might be realized for extreme values of the trend parameters a) in (6.45)] into (6.50 a, b), the conditions (6.51) lead to

$$\overline{E}_j < \infty \qquad E_{10}E_{20}(\gamma_{11} + \gamma_{12})(\gamma_{21} + \gamma_{22}) < 1 \tag{6.52}$$

and

$$\overline{E}_j > 0 \qquad \begin{aligned} &\nu_j + E_{i0}(1 - \nu_i)(\gamma_{jj} + \gamma_{ji}) < 1 && (6.53\,\text{a})\\ &\nu_j + E_{i0}(1 + \nu_i)(\gamma_{jj} + \gamma_{ji}) > -1 && (6.53\,\text{b})\end{aligned}$$

with $i, j = 1, 2$ where $i \neq j$. This implies that the relations (6.52, 53) have to be fulfilled by consistent parameter sets $\{E_{j0}, \nu_j, \gamma_{jj}, \gamma_{ji}\}$.

6.3.2 Numerical Analysis of Non-Stationary Solutions

Before discussing numerical, time dependent solutions of the model equations (6.28, 37, 39) for abstract interacting societies, a definite procedure for the selection of trend parameters has to be adopted. The parameters will be divided into two main sets where those of the first set are kept constant in all the computed (abstract) scenarios discussed here, while those of the second set will assume different values.

The values of the parameters of the first set are assumed to be

$$\left.\begin{aligned}
&\text{a)}\ \delta_{10} = \delta_{20} = 0\\
&\quad\ \alpha_{11} = \alpha_{22} = \alpha_{21} = 1.0\\
&\quad\ \beta_{11} = \beta_{22} = 2.0\\
&\quad\ E_{1S} = E_{2S} = 0.01\ ,\\
&\text{b)}\ \nu_1 = \nu_2 = 0.2\\
&\quad\ \gamma_{11} = \gamma_{22} = \gamma_{21} = \gamma_{12} = 0\\
&\quad\ \varepsilon = 10\\
&\quad\ E_{10} = E_{20} = 1.0\\
&\text{c)}\ P_{1\text{th}} = P_{2\text{th}} = -0.9\\
&\quad\ \mu_{11} = \mu_{22} = 0.5\\
&\quad\ \mu_{12} = \mu_{21} = 0.1\ .
\end{aligned}\right\} \tag{6.54}$$

The implications of these parameter choices are discussed below:

a) $\delta_{10} = \delta_{20} = 0$ implies that no "instrinsic" preference for dovishness or hawkishness with respect to the other society is assumed in the societies $\mathcal{S}_i$.

The values $\alpha_{11} = \alpha_{22} = \alpha_{21} = 1.0$ fix the levels of the constructive (destructive) influences of the internal (the competing) economic capacity – in terms of E_1 and E_2 – on the political psychology in $\mathcal{S}_1$ and $\mathcal{S}_2$ (α_{12} will be varied and therefore belongs to the second parameter set). The signs of α_{jj} and α_{ji} correspond to the competitive attitude between $\mathcal{S}_1$ and $\mathcal{S}_2$.
$\beta_{11} = \beta_{22} = 2.0$ determines the level of the negative reaction of P_j to any destruction inflicted on $\mathcal{S}_j$ by military measures of the rivaling society $\mathcal{S}_i$.
$E_{1S} = E_{2S} = 0.01$ defines the level of equipment (after a breakdown in the case of military conflict) at which war weariness and conciliatory attitudes appear and/or reappear.

b) $\nu_1 = \nu_2 = 0.2$ determines the extent of strengthening or weakening feedback effects of the internal political psychology on the economic development in $\mathcal{S}_j$. The parameters γ_{jj} and γ_{ji} describe a similar effect essentially reinforcing the effect of the parameters ν_j. Hence $\gamma_{11} = \gamma_{22} = \gamma_{21} = \gamma_{12} = 0$ can also be assumed.
$E_{10} = E_{20} = 1.0$ fixes the potential size of the economy (in terms of total equipment) of both systems in that E_{j0} is the stationary value of E_j if no interaction with other variables occurs. For "competitive" societies it is reasonable to assume equal potential levels E_{10} and E_{20}.
$\varepsilon = 10$ defines the time scale on which changes in the material (economic) variables occurs. The assumption that these changes occur typically one order of magnitude slower than those of political psychology seems reasonable.

c) The choice $P_{1\text{th}} = P_{2\text{th}} = -0.9$ implies strongly hawkish thresholds which, together with the values

$$\mu_{11} = \mu_{22} = 0.5 \qquad \text{and} \quad \mu_{12} = \mu_{21} = 0.1$$

lead to a de-escalating situation (according to (6.43 b)) for "normal" i.e. not extremely negative values of P_1 and P_2.

The second set consists of the trend parameters $\{\kappa_1, \kappa_2, \kappa_{12}, \kappa_{21}, \alpha_{12}, \beta_{12}, \beta_{21}\}$. Their values, together with varying conditions for the initial destruction rates, $D_1(0)$ and $D_2(0)$, and the values belonging to these choices, are listed in Table 6.1. Each coupled scenario is built up using the constant information discussed above in connection with one row of parameter values from Table 6.1.

Before going into further details it should be stressed that only *abstract scenarios* for *abstract societies* $\mathcal{S}_1$ and $\mathcal{S}_2$ will be discussed since a description of fully realistic situations as opposed to certain typical structures would certainly overburden the minimal model considered here.

The Figs. 6.2–16 present the resulting coupled paths of the variables $E_j(\tau)$ (total equipment of $\mathcal{S}_j$), $P_j(\tau)$ (political psychology in $\mathcal{S}_j$) and $D_j(\tau)$ (destruction rate in $\mathcal{S}_j$) in scaled time τ for the values (6.54) of the first set of trend parameters together with the values listed in Table 6.1 for the second set. The paths for society $\mathcal{S}_1$ are shown as solid lines and those for $\mathcal{S}_2$ dashed. While the initial conditions $D_j(0)$ are varied from case to case, the same initial conditions $E_1(0) = 0.2$, $E_2(0) = 0.5$ and $P_1(0) = -0.2$, $P_2(0) = 0$ apply to all cases discussed. Hence, society $\mathcal{S}_1$ starts from a poorer equipment level than $\mathcal{S}_2$ despite

Table 6.1. Parameter values corresponding to the scenarios presented in Figs. 6.2–16

Group	Parameter Set	κ_1	κ_2	κ_{12}	κ_{21}	α_{12}	β_{12}	β_{21}	$D_1(0)$	$D_2(0)$
1	6.2	0.8	0.8	2.0	0.0	1.0	0.0	0.0	0.0	0.0
	6.3	0.8	0.8	2.0	0.0	1.0	0.0	0.0	0.0	0.18
	6.4	0.8	0.8	2.0	0.0	1.0	0.0	0.0	0.0	0.30
2	6.5	1.0	1.0	2.0	0.0	1.0	0.0	0.0	0.0	0.0
	6.6	1.0	1.0	2.0	0.0	1.0	0.0	0.0	0.0	0.18
	6.7	1.0	1.0	2.0	0.0	1.0	0.0	0.0	0.0	0.30
3	6.8	1.0	1.0	2.0	0.0	3.0	0.0	0.0	0.0	0.0
	6.9	1.0	1.0	2.0	0.0	2.0	0.0	0.0	0.0	0.0
	6.10	1.0	1.0	2.0	0.0	0.0	0.0	0.0	0.0	0.0
4	6.11	1.0	1.0	0.5	0.5	1.0	0.0	0.0	0.0	0.0
	6.12	1.0	1.0	0.5	0.5	1.0	0.0	0.0	0.0	0.095
	6.13	1.0	1.0	0.5	0.5	1.0	0.0	0.0	0.0	0.30
	6.14	1.0	1.0	0.5	0.5	1.0	0.0	0.0	0.0	0.50
	6.15	1.0	1.0	0.5	0.5	1.0	0.02	0.02	0.0	0.30
	6.16	1.0	1.0	0.5	0.5	1.0	0.0	0.0	0.50	0.0

the fact that the same "potential levels" $E_{01} = E_{02} = 1$ exist. In the cases when "permanent peace" occurs, i.e. $D_1(\tau) = D_2(\tau) = 0$, only $E_j(\tau)$ and $P_j(\tau)$ are presented (Figs. 6.2, 5, 8–11). The parameter values in Table 6.1, each set of which represents a certain scenario to be investigated, are arranged in four groups and the progression from one group to the next involves a change in only one trend parameter type.

Group 1: *Figures* 6.2–4. This group of scenarios is characterized by a relatively weak internal political opinion pressure in both societies [$\kappa_1 = \kappa_2 = 0.8$ is below the threshold value $\kappa_c = 1.0$ of a transition from a monostable to a bistable potential $V_j(P_j, \ldots)$ – see (6.29 a, b)]. Furthermore, because $\kappa_{12} = 2$ the political psychology P_1 of $\mathcal{S}_1$ is strongly coupled to the corresponding political climate P_2 in $\mathcal{S}_2$ while, since $\kappa_{21} = 0$, the reverse is not true.

In Fig. 6.2 the initial condition $D_1(0) = D_2(0) = 0$ leads to a permanently peaceful situation. The material variables $E_1(\tau)$ and $E_2(\tau)$ soon reach the saturation level 1.0 and afterwards perform small oscillations around this level sustained by the interaction between the economic and political variables. The variables of political psychology oscillate at higher amplitudes since they react strongly to economic disequilibria; the oscillation is further enhanced in $\mathcal{S}_1$ because of the more sensitive reaction of P_1 to P_2.

In Fig. 6.3 the initial conditions are changed by adopting a military provocation of $\mathcal{S}_2$ by $\mathcal{S}_1$, represented by a relatively low initial destruction rate $D_2(0)$. Although this provocation elicits a weak military reaction of $\mathcal{S}_2$ and minor distortions of the initial paths of the economic and political variables in $\mathcal{S}_1$ and $\mathcal{S}_2$, the two systems still remain in the domain of de-escalation. Since $P_1(\tau)$ and $P_2(\tau)$ stay far above the threshold values P_{1th} and P_{2th}, the condition (6.43 b)

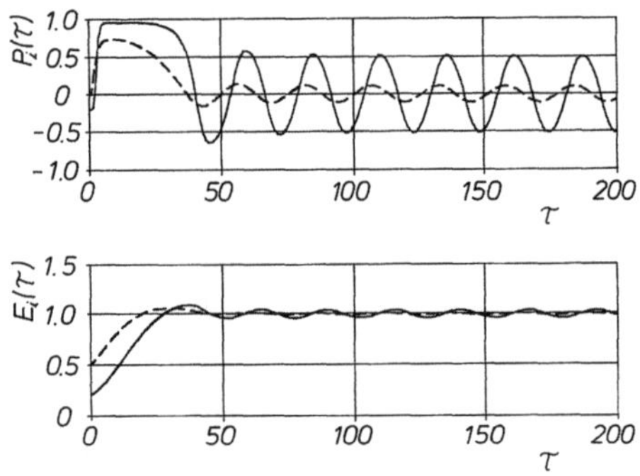

Fig. 6.2. Evolution of the variables $E_i(\tau)$ and $P_i(\tau)$ in $\mathcal{S}_1$ (—) and $\mathcal{S}_2$ (---) for trend parameters listed in (6.54) and row (6.2) of Table 6.1

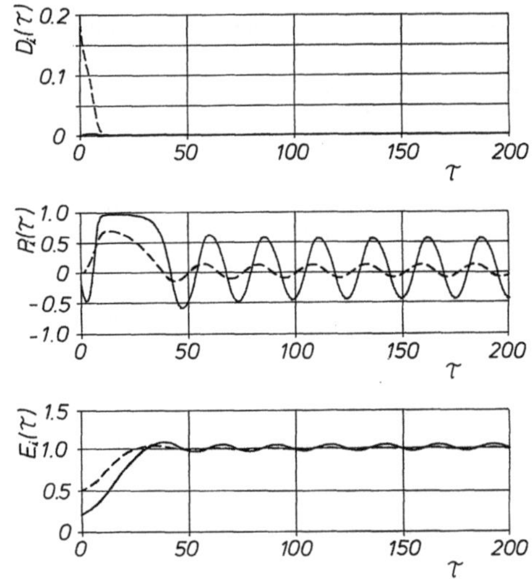

Fig. 6.3. Evolution of the variables $E_i(\tau)$, $P_i(\tau)$ and $D_i(\tau)$ in $\mathcal{S}_1$ (—) and $\mathcal{S}_2$ (---) for trend parameters listed in (6.54) and row (6.3) of Table 6.1

expressing the dominance of conciliatory attitudes remains fulfilled and destruction pulses decay. After the provocation is settled the eco-political cycles similar to those occurring under peaceful conditions, as in Fig. 6.2 set in.

In Fig. 6.4 the sequence of events is dominated by the assumed military agression of $\mathcal{S}_1$ (high initial destruction rate $D_2(0)$ inflicted on $\mathcal{S}_2$). In comparison to Fig. 6.3 a completely different scenario consisting of characteristic periods develops. In the period:

$0 < \tau < 30$: the initial destruction in $\mathcal{S}_2$ leads to an extreme mutually reinforcing "hawkishness" $P_1 \lesssim P_{1\text{th}}$ and $P_2 \lesssim P_{2\text{th}}$ in both systems followed by an escalation of the destructiveness variables Q_1 and Q_2 and an ensuing continuation of mutual destruction, i.e. a war occurs. The destruction generates a breakdown of the equipment variable for the society $\mathcal{S}_2$, i.e. $\mathcal{S}_2$ is the victim of the initial aggression.

$30 < \tau < 40$: the material breakdown in $\mathcal{S}_2$ elicits a strong mutual climate of reconciliation leading to a definite end to destruction. The first post-war period begins.

$40 < \tau < 50$: the marked economic disequilibrium between $E_1(\tau)$ and $E_2(\tau)$, a remnant of the war, drags both societies into new political tension; this is the second post-war period.

$50 < \tau < 120$: the slow recovery of the economy of $\mathcal{S}_2$ relative to that of $\mathcal{S}_1$ finally leads to a corresponding improvement in the mutual political climate. The third post-war period ends.

$\tau > 120$: "peace-time oscillations" of the economy and the political relationship set in, showing the same pattern as the end conditions in Figs. 6.2, 6.3.

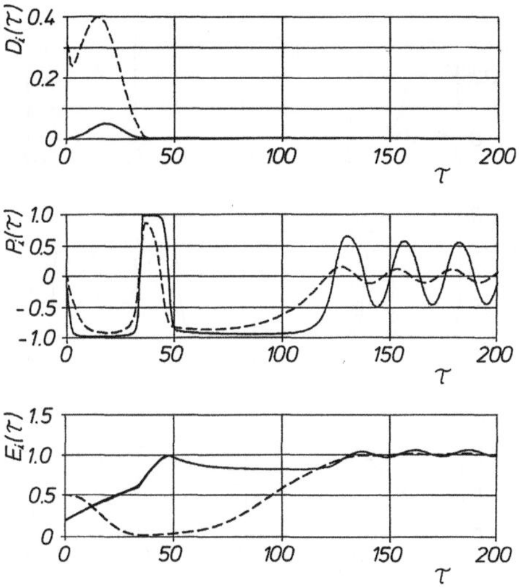

Fig. 6.4. Evolution of the variables $E_i(\tau)$, $P_i(\tau)$ and $D_i(\tau)$ in $\mathcal{S}_1$ (——) and $\mathcal{S}_2$ (---) for trend parameters listed in (6.54) and row (6.4) of Table 6.1

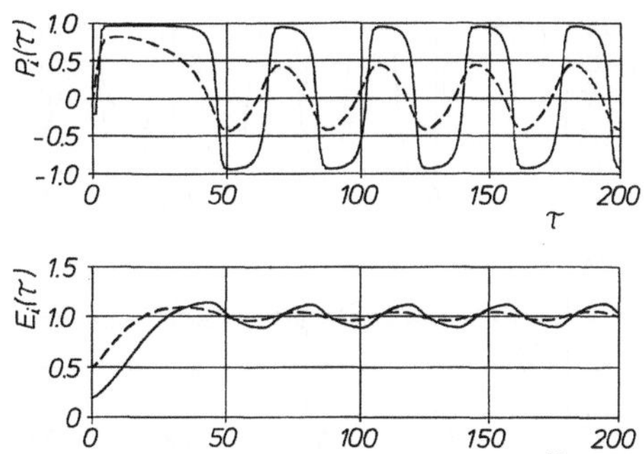

Fig. 6.5. Evolution of the variables $E_i(\tau)$ and $P_i(\tau)$ in $\mathcal{S}_1$ (——) and $\mathcal{S}_2$ (---) for trend parameters listed in (6.54) and row (6.5) of Table 6.1

Group 2: *Figures* 6.5–7. The second group of figures is obtained by changing only one feature of the competing systems; a strong internal political opinion pressure (expressed by the values $\kappa_1 = \kappa_2 = 1.0$) is now assumed. Comparing Figs. 6.5–7 with the corresponding Figs. 6.2–4, it can be observed that the amplitudes of the political climate oscillations are enhanced – because of their higher collective impetus – which also induces larger economic fluctuations. The structures of the scenarios, however, remain the same and can be interpreted along the same lines as those of Group 1.

Group 3: *Figures* 6.8–10. In this group two societies in asymmetrical eco-political, although peaceful non-destructive relationship are considered (Table 6.1). Because $D_1(\tau) = D_2(\tau) = 0$, the variables $D_j(\tau)$ need not be taken into account. The eco-political asymmetry comes about through the parameter choices $\kappa_{12} = 2.0$ and $\kappa_{21} = 0.0$ – as in Groups 1 and 2 – and $\alpha_{12} = 3.0, 2.0, 0.0$ (while $\alpha_{12} = 1.0$ belongs to the constant parameters of the first set). Parameter choices $\kappa_{12} > \kappa_{21}$, $\alpha_{12} > \alpha_{21}$ and an initial condition $E_1(0) < E_2(0)$ imply that society $\mathcal{S}_1$ is eco-politically "weaker" than society $\mathcal{S}_2$ in the following sense: The economic variable E_1 starts from a lower level than E_2; the political climate P_1 in $\mathcal{S}_1$ reacts sensitively to that of $\mathcal{S}_2$ but not vice versa; furthermore, the competition between $\mathcal{S}_2$ and $\mathcal{S}_1$ leads to a stronger negative reaction of the climate P_1 to economic prosperity in $\mathcal{S}_2$ than occurs vice versa. The political climate of $\mathcal{S}_1$ and $\mathcal{S}_2$, on the other hand, influences the respective economical situations.

Figure 6.8 shows that the strong negative influence of the economy E_2 of $\mathcal{S}_2$ on the climate P_1 of $\mathcal{S}_1$ ($\alpha_{12} = 3.0$) leads to a stationary separation of $\mathcal{S}_2$ and $\mathcal{S}_1$. The society $\mathcal{S}_2$ develops into an economically prosperous and politically positive

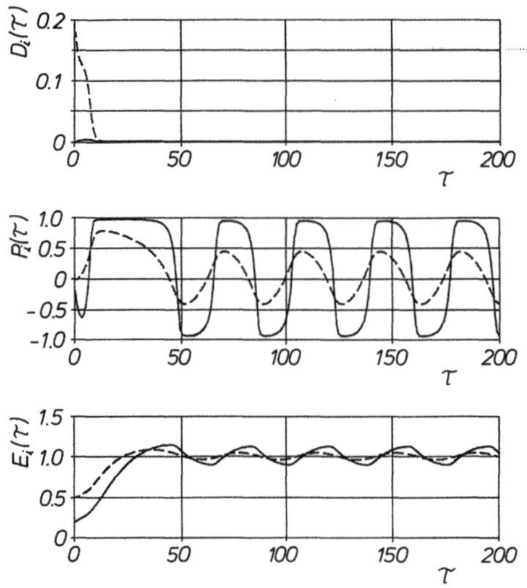

Fig. 6.6. Evolution of the variables $E_i(\tau)$, $P_i(\tau)$ and $D_i(\tau)$ in $\mathcal{S}_1$ (—) and $\mathcal{S}_2$ (---) for trend parameters listed in (6.54) and row (6.6) of Table 6.1

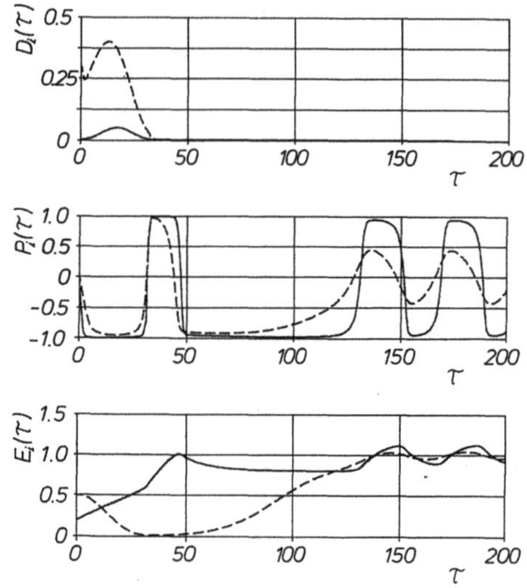

Fig. 6.7. Evolution of the variables $E_i(\tau)$, $P_i(\tau)$ and $D_i(\tau)$ in $\mathcal{S}_1$ (—) and $\mathcal{S}_2$ (---) for trend parameters listed in (6.54) and row (6.7) of Table 6.1

minded, "optimistic" state, while the society $\mathcal{S}_1$ remains in an economically depressed and politically adverse state.

Figure 6.9 shows a milder, though with $\alpha_{12} = 2.0$ and $\alpha_{21} = 1.0$ still asymmetrical competition between the societies $\mathcal{S}_1$ and $\mathcal{S}_2$. Both societies are now economically prosperous although $\mathcal{S}_2$ is slightly better off. Because of sensitive reactions in $\mathcal{S}_1$ to events in $\mathcal{S}_2$ the political climate fluctuations in $\mathcal{S}_1$ are more pronounced than those of $\mathcal{S}_2$.

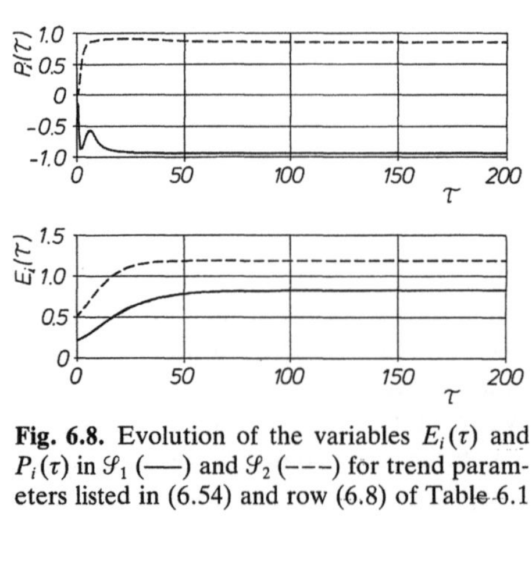

Fig. 6.8. Evolution of the variables $E_i(\tau)$ and $P_i(\tau)$ in $\mathcal{S}_1$ (—) and $\mathcal{S}_2$ (---) for trend parameters listed in (6.54) and row (6.8) of Table 6.1

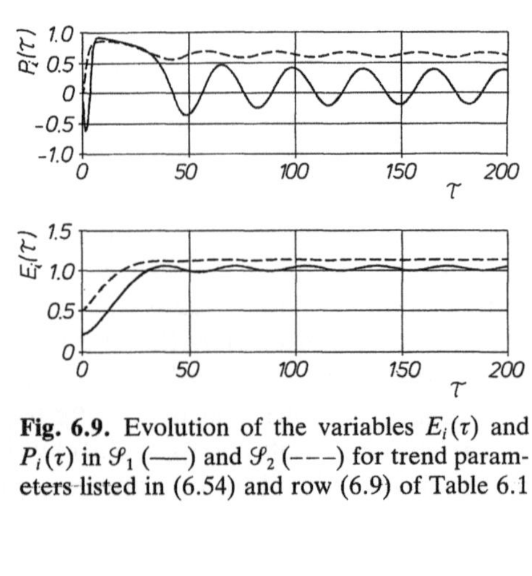

Fig. 6.9. Evolution of the variables $E_i(\tau)$ and $P_i(\tau)$ in $\mathcal{S}_1$ (—) and $\mathcal{S}_2$ (---) for trend parameters listed in (6.54) and row (6.9) of Table 6.1

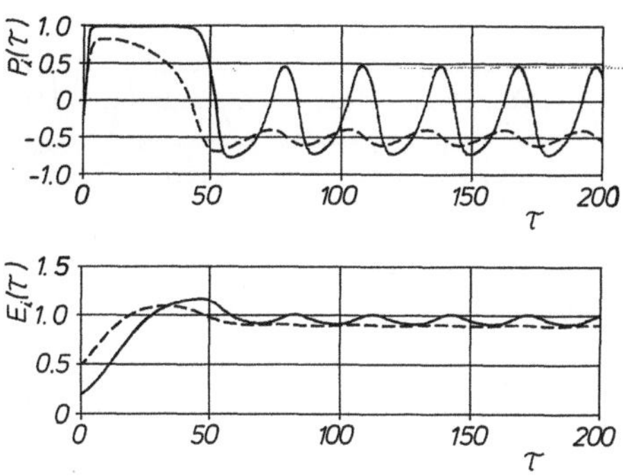

Fig. 6.10. Evolution of the variables $E_i(\tau)$ and $P_i(\tau)$ in $\mathcal{S}_1$ (—) and $\mathcal{S}_2$ (---) for trend parameters listed in (6.54) and row (6.10) of Table 6.1

In Fig. 6.10 the asymmetry of economic influences on the political psychology has been reversed by choosing $\alpha_{12} = 0$ and $\alpha_{21} = 1.0$. The economic prosperity of both systems has decreased on average and society $\mathcal{S}_1$, though fluctuating in its political climate, is now economically slightly better off than $\mathcal{S}_2$.

Group 4: *Figures* 6.11–16. In this group interacting societies with completely symmetrical trend parameters are considered. Strong internal opinion pressures, i.e. high propaganda levels ($\kappa_1 = \kappa_2 = 1.0$), moderate mutual political "climate-exchanges" ($\kappa_{12} = \kappa_{21} = 0.5$) and competitive eco-political interactions ($\alpha_{12} = \alpha_{21} = 1.0$) are assumed. However, the initial conditions are chosen as asymmetrical.

Figure 6.11 shows how – despite peaceful conditions $D_1(\tau) = D_2(\tau) = 0$ – a permanent negative situation can stabilize between symmetrical systems. By mutually imposing a stationary adverse political climate on themselves both economies remain – because of destructive interaction via the negative political climate – below their potential maximum levels.

The stationarity of the solution in Fig. 6.11 does not lead to the conclusion that such systems always tend to stability. In fact the contrary is true, as observed in Figures 6.12–14, where initial military action against $\mathcal{S}_2$ (i.e., initial destruction rates of different magnitudes in $\mathcal{S}_2$) are assumed. These initial conditions unleash a regime of permanent crisis demonstrating the extreme instability of the phase of destruction and warfare.

Considering for instance the scenario of Fig. 6.12, the following sequence of events is observed:

$0 < \tau < 240$: The initial attack by $\mathcal{S}_1$ against $\mathcal{S}_2$ leads to retaliation in the form of two counter attacks by $\mathcal{S}_2$ against $\mathcal{S}_1$; simultaneously the political climate in both societies passes the threshold value eliciting the escalation of destructiveness; economic equipment is destroyed to a large extent.

$240 < \tau < 260$: War weariness leads to a conciliatory climate in $\mathcal{S}_2$ which corresponds to a cessation in destructive action against $\mathcal{S}_1$. This concession on the part of $\mathcal{S}_2$, however, is not rewarded by corresponding reaction in the society $\mathcal{S}_1$ which remains politically militant and continues its military action against $\mathcal{S}_2$.

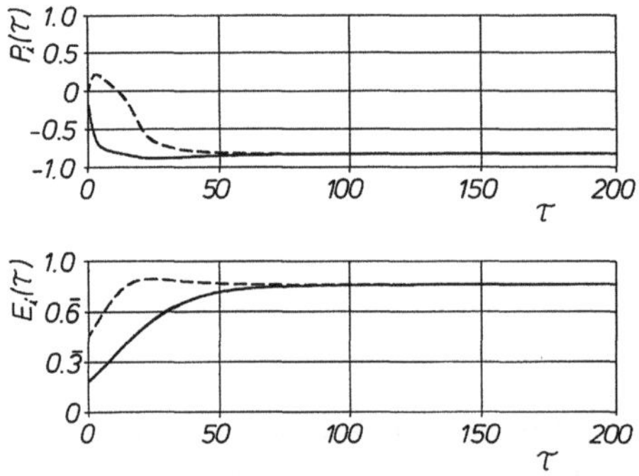

Fig. 6.11. Evolution of the variables $E_i(\tau)$ and $P_i(\tau)$ in $\mathcal{S}_1$ (—) and $\mathcal{S}_2$ (---) for trend parameters listed in (6.54) and row (6.11) of Table 6.1

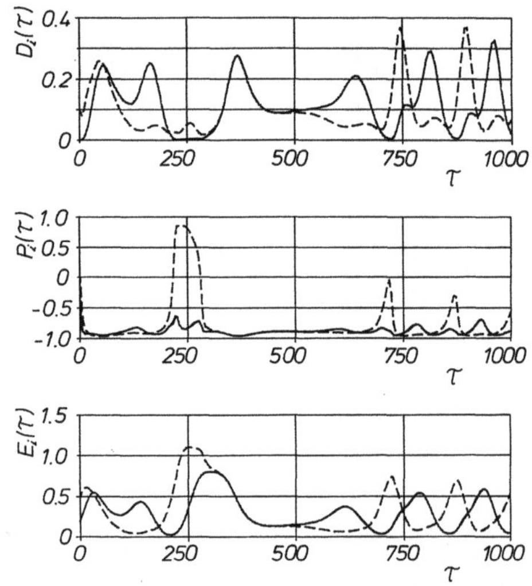

Fig. 6.12. Evolution of the variables $E_i(\tau)$, $P_i(\tau)$ and $D_i(\tau)$ in $\mathcal{S}_1$ (—) and $\mathcal{S}_2$ (---) for trend parameters listed in (6.54) and row (6.12) of Table 6.1

$260 < \tau < 500$: The ensuing failure to take advantage of this opportunity for peace leads to another period of militant climate, destruction and depression of the economic state in both societies, events, which, however, are not radical enough to elicit a definite war weariness and a sufficiently conciliatory climate to enable an end to military action.

$500 < \tau < 1000$: At $\tau = 500$ the variables of both societies have almost the same values, i.e. $E_1 \approx E_2$, $P_1 \approx P_2$ and $D_1 \approx D_2$; by assumption, all trend parameters of $\mathcal{S}_1$ and $\mathcal{S}_2$ are also equal. This means that the equations have a symmetrical structure with respect to the societies $\mathcal{S}_1$ and $\mathcal{S}_2$. In spite of this the nonlinear structure of the equations leads to widely differing trajectories for $\mathcal{S}_1$ and $\mathcal{S}_2$ in the macrovariable $\{E, P, D\}$-space in the following period. This period is again one of permanent crisis filled with attacks, counterattacks and short periods of reconstruction of the respective economies. It seems that the societies $\mathcal{S}_1$ and $\mathcal{S}_2$ can only be released from their mutually sustained crisis situation through the mediating efforts of third societies, whose interaction with $\mathcal{S}_1$ and $\mathcal{S}_2$, of course, is not described in the model.

The macrovariable paths of the Figs. 6.13, 14 have to be interpreted on the same lines as those of Fig. 6.12. Whereas the results shown in Figs. 6.12–14 could lead to the pessimistic conclusion, that – at least under the built-in model conditions – crisis states including mutual destruction can be of an irregularly fluctuating permanent nature, the last Figs. 6.15, 16 in the series of model runs show, that certain – more or less accidental – shifts in trend parameters or initial conditions may open channels leading to an end of the war. In Fig. 6.15, where

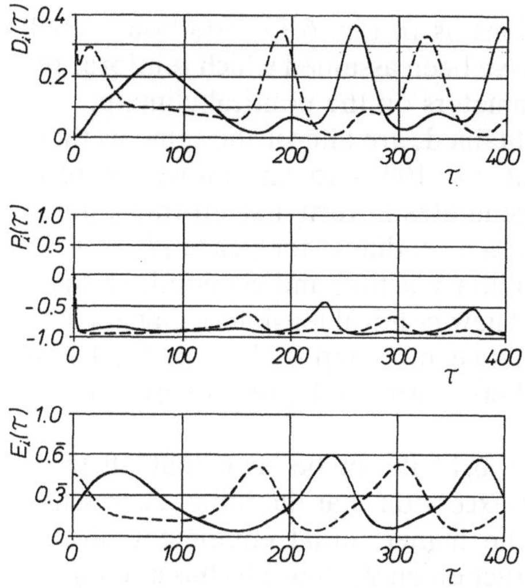

Fig. 6.13. Evolution of the variables $E_i(\tau)$, $P_i(\tau)$ and $D_i(\tau)$ in $\mathcal{S}_1$ (—) and $\mathcal{S}_2$ (---) for trend parameters listed in (6.54) and row (6.13) of Table 6.1

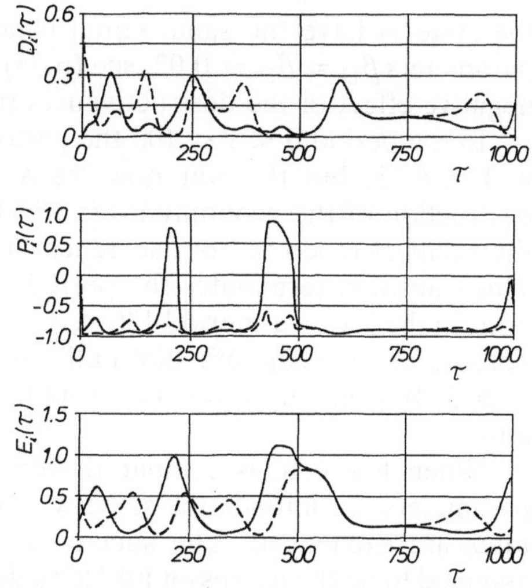

Fig. 6.14. Evolution of the variables $E_i(\tau)$, $P_i(\tau)$ and $D_i(\tau)$ in $\mathcal{S}_1$ (—) and $\mathcal{S}_2$ (---) for trend parameters listed in (6.54) and row (6.14) of Table 6.1

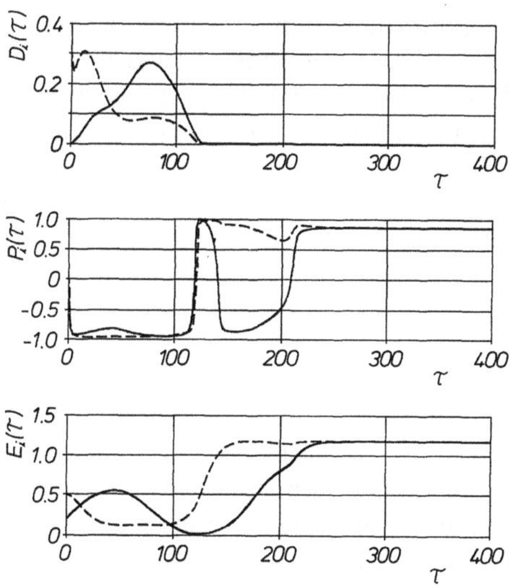

Fig. 6.15. Evolution of the variables $E_i(\tau)$, $P_i(\tau)$ and $D_i(\tau)$ in $\mathcal{S}_1$ (—) and $\mathcal{S}_2$ (---) for trend parameters listed in (6.54) and row (6.15) of Table 6.1

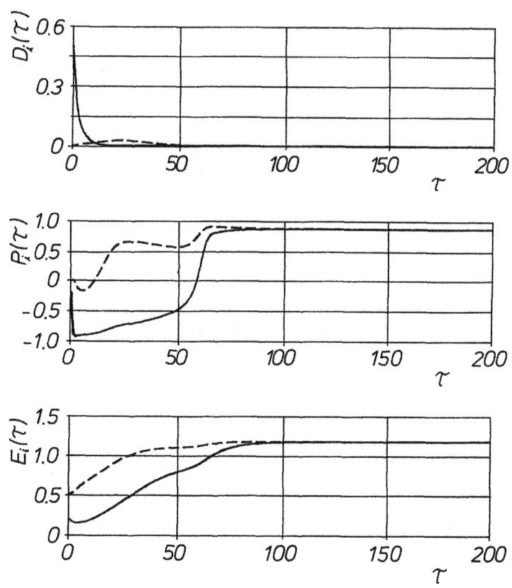

Fig. 6.16. Evolution of the variables $E_i(\tau)$, $P_i(\tau)$ and $D_i(\tau)$ in $\mathcal{S}_1$ (—) and $\mathcal{S}_2$ (---) for trend parameters listed in (6.54) and row (6.15) of Table 6.1

the systems have the same initial conditions as in Fig. 6.13, additional small parameters $\beta_{12} = \beta_{21} = 0.02$, see (6.35), have been assumed which reinforce the negative effect of the destructiveness parameters on the political climate.

In the period $0 < \tau < 100$ the paths obtained, are almost the same as those in Fig. 6.13, but the war now leads – at $\tau \approx 100$ – to an almost complete destruction of the economy in $\mathcal{S}_1$ which generates in turn a steep transition in the political mood in both societies towards a readiness for peace ($P_j \Rightarrow +1$). This transition terminates the period of military action and economic destruction. In the post-war period $130 < \tau < 210$ new political tension reappears in $\mathcal{S}_1$ because of the economic disequilibrium between $\mathcal{S}_2$ and $\mathcal{S}_1$ but the final state ($\tau > 210$) is one of stationary eco-political prosperity and, presumably, cooperation.

When Fig. 6.16 is compared with Fig. 6.14 it can be seen that all trend parameters are initially the same with the exception that the initial destruction rates are exchanged. The society with the higher initial equipment level is assumed to be the aggressor inflicting destruction on $\mathcal{S}_1$. Since $\mathcal{S}_1$ has a relatively weak capacity to strike back, the political climate in $\mathcal{S}_2$ is not completely spoiled and leads to a rather rapid de-escalation in destructive action; the subsequent development of $\mathcal{S}_1$ lags behind that of $\mathcal{S}_2$ but finally both societies reach the same prosperous eco-political level.

Reviewing the selection of scenarios presented in this section, one summarizing remark should be made immediately. The extreme sensitivity of the dynamics to the initial conditions and numerical values of the trend parameters in unstable situations prohibits any use of models of the type discussed here for predictive purposes in such situations. Critical conditions giving rise to an all-important role of "micro-fluctuations" (such as those that can be induced by prominent politicians setting the – positive or negative – direction and trend of future development at points of instability) cannot be modelled using macrovariables only, i.e. the microscopic theory of direction and timing of such fluctuations is certainly beyond the scope of models working only with grossvariables. But such models *can* give an account of the *domains of critical dynamic behaviour* of the macrosystems investigated.

6.4 Political Implications

Finally, the question of the political implications of the foregoing results should be discussed. The initial observation that variables of political psychology, comprising both rational and emotional or irrational components of the ideologies of macrosocieties, play a decisive role in the dynamics of these societies is a reasonably obvious one. Deeper lying reasons for the fact that ideologies support this kind of dynamics are their indispensibility and their ambiguity. Because they provide ordering principles and guidelines for societies, ideologies are *indispensible*. On the other hand, no ideology disposes over the "absolute truth" or "absolute values"; in fact an ideology is *ambiguous* by its very nature since it provides at best a model for the relative optimization of the coexistence of competing ideas on socio-economic structures and ways of life in the macrosociety in which it exists.

The necessity for and, on the other hand, the ambiguity or relativity of ideologies has political implications for the interaction of macrosocieties.

From one point of view an ideology has the inherent right of self-assertion and of evolution within and according to the direction given to it by the society in which it exists. From another point of view no ideology has any natural right to dominate *all* societies, for instance, by claiming absolute validity or by striving for an ultimate "victory" over other ideologies.

It is obvious that this "meta-ideological" moral constraint on the competition of macrosocieties becomes the more important, the more speculation about unleashing a phase transition to war as a means of "deciding" between ideologies is considered as being morally absolutely forbidden.

One may ask, whether in this situation any definite conclusions on the dynamics of macrosocieties (for instance, by considering general properties or, more specifically, the role of certain trend parameters in a model) can be derived from the quantified theory presented here. Of course, no new insight on principles can be expected, but arguments on issues of policy for macrosocieties can be provided in terms of the quantitative theory.

One general observation is that the dynamics of sociological systems typically obey *non-linear equations of motion*. Such equations not only describe the more trivial case of a development towards stationary states, but also a variety of phase transitions and of transient states. Because circumstances *do* change in the long term it seems unavoidable that the global development of the human society also includes transient behaviour. However, different channels of development with either prevailingly "soft" or prevailingly "hard" phase transitions are feasible.

In view of the uncontrollable, perhaps irreversible and devastating effects of radical phase transitions a world-wide policy influencing trend parameters in a way that leads to controllable, calculable smooth and soft phase transitions instead of to hard ones has to be preferred from the standpoint of political morale. The question then is: What could facilitate and what hinder smooth phase transitions?

At first some insiduous properties of nonlinear dynamics characterizing the difficulties of finding the intended smooth channels of development have to be considered. For *linear dynamics* the "effects" are, in a rather general sense, proportional to the "causes" (forces, trends, restraints). On the other hand, within a *non-linear dynamics* the global effects need not be proportional to the strength and direction of the eliciting forces. Under critical circumstances the magnitude and direction of the effects of certain forces is practically uncontrollable. In particular, counterproductive effects are possible, if not probable. So a moral policy should be to try to avoid critical situations in which the total effect of all actions becomes unpredictable.

The two extreme opposite attitudes in competitive macrosocieties likely to produce counteractive effects must be – at least hypothetically – discussed: The one attitude vigorously advocates the destabilisation of the ideology and the material basis of the competing society using all means available. This attitude may elicit either aggressive counteractions or a sudden internal destabilisation of the competing society, and both of these are rapid, non-calculable transitions.

The other attitude vigorously advocates the "direct way to peace", for instance by sudden one-sided disarmament of the own society. The resulting disequilibrium may also elicit counteractive destabilizing effects, for instance, the temptation occurring in the other society to take advantage of the situation in a way leading to new non-calculable risks.

What, then, could be general rules for producing smooth instead of explosive transient states under the difficult conditions of non-linear dynamics? The models show (see, for instance, those of Chaps. 2 and 6) that strong trend parameters of collective formation, which lead to a strong dependence of individual transition probabilities on a few collective order parameters, generate a rigid system which may undergo sudden destabilizing transitions into other phases. It can be concluded that:

Avoiding the danger of *sudden and explosive internal transitions* amounts to slowly and "adiabatically" introducing a more open and flexible instead of a more rigid and dogmatic ideology; furthermore it amounts to evading less controllable collective emotionalisation arising, for instance, from biased or ideologized information, from an accumulation of problems or from postponing necessary open discussion, or reform, etc.

Avoiding sudden external transitions and adverse interactions between macrosocieties amounts to building-up at least a minimal degree of cooperation and mutual trust in order to reach a safe, instead of escalatory modus of competition and therefore a state of coexistence on all material and ideological levels. Certainly frank mutual criticism could still occur but the idea of destabilizing the very substance of the competing system not only for preliminary tactical reasons but permanently must be renounced.

Only if the internal dynamics *within* the macrosocieties and the external dynamics *between* them, including their difficulties and possible instabilities, are understood and the problems arising are regarded as problems to be solved through cooperation *between the societies*, can there exist a chance of solving the present fundamental coexistence problem of mankind: The aim has to be a substitution of the traditional disastrous dynamics of quasi-periodic transitions between phases of peace, tension and war by a new smooth and evolutionary dynamics avoiding phases of rapid and too often catastrophic change.

While considering the present state of the coexistence problem in terms of the concepts of quantitative sociology, it can be observed that mankind still lives in a state of relatively underdeveloped meta-ideological reasoning. As a consequence the roman dictum "Si vis pacem, para bellum" is still in operation. It presently takes the form of the doctrine of the attainment of a relative safety through mutual deterrence: Weapons of deterrence generate trend parameters for the political psychology leading to a safe domain for the paths of the macrovariables of both societies. However, adjacent to the domain of relative safety a domain of high instability is found in which all paths end in devastated or chaotic states. This situation may be illustrated by thinking of a mountaineer going astray in a highland valley of relative safety surrounded by deadly precipices. The danger is increased by the fact that the essential ingredients of the situation – mutual fear, ideological subversion, eventual breakthroughs in the development of new weapons – lead to unstable trend parameters. Thus the shape of the domain of relative safety itself may vary since this is determined by the trend parameters of the equations of motion of the macrovariables; for example, the threshold of warfare can be shifted by the introduction of new kinds of weapons.

This picture, which is based only on the general features of the non-linear dynamics of societies, leads to conclusions which could reconcile opposing, or lend substance to controversial arguments occurring in the discussion on the preservation of peace: A *short term strategy* with this aim can only consist of a conservation of the present domain of relative safety by the above mentioned means, because any sudden changes in trend parameters – relatively independent of their direction and of the motivations behind them – will deform the present safe domain in an (almost) unpredictable way with effects that are practically not calculable. On the other hand, a *long term strategy* has to be of a different nature: It should consist of the slow substituation of the unstable trend parameters of military deterrence and extreme ideological rivalry by the more stable trend parameters of meta-ideological reason! This "adiabatical" substitution if controlled by mutual agreement between the competing societies can lead in the end to an enlarged and more stable safe domain of development for both macrosocieties.

The process of transition from the short term to the long term strategy in generating enlarged safe domains of development will turn out to be a highly delicate task because it presumes the cooperation of the competing macrosocieties, and is therefore highly unstable with respect to errant behaviour of one of the partners. The success in finding this transition path to safer domains will depend on the wisdom of the statesmen and people of both societies in handling the trend parameters of the complex dynamics of political psychology.

Quantitative sociology – being meta-ideological in its nature – can only describe and try, in a positive sense, to accompany such a process, for instance by substituting the old dictum "Si vis pacem para bellum" with "Si vis pacem, para exemplum de bello non gerendo".

References and Further Reading

It is not possible here to give a complete list of references for further general reading. It is only possible to quote a selection of books and articles which might be useful for a general introduction into the fields mentioned or for special studies of problems closely related to the text.

Chapter 1

1.1 H. Haken: Cooperative phenomena in systems far from thermal equilibrium and in nonphysical systems. Rev. Mod. Phys. **47,** 67 (1975)

1.2 H. Haken (ed.): *Cooperative Effects, Progress in Synergetics,* (North-Holland, Amsterdam 1974)

1.3 H. Haken: *Synergetics, an Introduction,* 2nd ed., Springer Ser. Synergetics, Vol. 1 (Springer, Berlin, Heidelberg, New York 1977)

1.4 H. Haken (ed.): *Synergetics, a Workshop*, Springer Ser. Synergetics, Vol. 2 (Springer, Berlin, Heidelberg, New York 1977)

1.5 A. Pacault, C. Vidal (eds.): *Synergetics far from Equilibrium,* Springer Ser. Synergetics, Vol. 3 (Springer, Berlin, Heidelberg, New York 1978)

1.6 W. Güttinger, H. Eikemeier (eds.): *Structural Stability in Physics,* Springer Ser. Synergetics, Vol. 4 (Springer, Berlin, Heidelberg, New York 1978)

1.7 H. Haken (ed.): *Pattern Formation by Dynamic Systems and Pattern Recognition,* Springer Ser. Synergetics, Vol. 5 (Springer, Berlin, Heidelberg, New York 1979)

1.8 H. Haken (ed.): *Dynamics of Synergetic Systems,* Springer Ser. Synergetics, Vol. 6 (Springer, Berlin, Heidelberg, New York 1980)

1.9 G. Nicolis, I. Prigogine: *Selforganisation in Non-Equilibrium System* (Wiley, New York 1977)

1.10 H. Haken: Generalized Ginzburg-Landau equations for phase transition-like phenomena in lasers, nonlinear optics, hydrodynamics and chemical reactions. Z. Phys. **B 21**, 105 (1975)

1.11 R. Bonifacio (ed.): *Dissipative Systems in Quantum Optics,* Topics Current Phys., Vol. 27 (Springer, Berlin, Heidelberg, New York 1982)

1.12 H. Haken: *Laser Theory,* Encyclopedia of Physics, Vol. XXV/2 (Springer, Berlin, Heidelberg, New York 1970)

1.13 F. Schlögl: Chemical reaction models for non-equilibrium phase transitions. Z. Phys. **253**, 147 (1972)

1.14 H. K. Janssen: Stochastic reaction model for a non-equilibium phase transition. Z. Phys. **270**, 67 (1974)

1.15 S. Chaturvedi, C. W. Gardiner, I. S. Matheson, D. F. Walls: Stochastic analysis of a chemical reaction with spatial and temporal structures. J. Stat. Phys. **17**, 469 (1977)

1.16 Y. Kuramoto: "Diffusion Induced Chemical Turbulence", in [Ref. 1.8, p. 134] and Progr. Theor. Phys. Suppl. **64**, 346 (1978)

1.17 A. Nitzan: „The Critical Behavior of Non-Equilibrium Transitions in Reacting Diffusion Systems", in [Ref. 1.8, p. 119]

1.18 M. Eigen, P. Schuster: The hypercycle, a principle of natural self-organization, Naturwissenschaften **64**, 541 (1977), **65**, 7 (1978), **65,** 341 (1978)

1.19 H. Kuhn: "Persistence of Organized Structures and Breakthrough of New Structures", in [Ref. 1.4, p. 200]

1.20 A. Babloyantz, L. K. Kaczmarek: Selforganization in multiply connected systems. Bull. Math. Biol. **41**, 193 (1979)
1.21 J. S. Nicolis, E. N. Protonotarios: Bifurcation in non-negotiable games, a paradigm for self-organization in cognitive systems. Int. J. Bio-med. Comput. **10**, 417 (1979)
1.22 J. S. Nicolis, M. Benrubi: Inadequate Communication Between Selforganizing Systems and Desynchronization of Physiological Rhythms, a Psychophysiological Study, General Systems, XXII, 119 (1977)
1.23 A. Gierer, H. Meinhardt: Adaptation and application of a theory of biological pattern formation. J. Cell. Sci. **15**, 321 (1974)
1.24 S. K. Ma: *Modern Theory of Critical Phenomena*, (Benjamin, New York 1976)
1.25 C. Prins (ed.): *Van der Waals Centennial Conference on Statistical Mechanics,* (North-Holland, Amsterdam 1974)
1.26 H. Haken: *Licht und Materie,* Vol. 1–3 (B. I.-Wissenschaftsverlag, Mannheim, Wien, Zürich, 1979, 1980, 1981)
1.27 H. Paul: *Lasertheorie,* Vol. 53, 54 (Akademie, Berlin WTB, 1969)
1.28 M. Sargent, M. O. Scully, W. E. Lamb, jr.: *Laser Physics,* (Addison Wesley, Reading, MA 1974)
1.29 F. Haake: *Treatment of Open Systems by Generalized Master Equations,* Springer Tracts Mod. Phys., Vol. 66 (Springer, Berlin, Heidelberg, New York 1973)
1.30 H. Grabert: *Projection Operator Techniques in Nonequilibrium Statistical Mechanics,* Springer Tracts Mod. Phys., Vol. 95 (Springer, Berlin, Heidelberg, New York 1982)
1.31 R. Landauer: The Role of Fluctuations in Multistable Systems and in the Transition to Multistability, in *Bifurcation Theory and Applications in Scientific Disciplines,* ed. by O. Gurel, O. E. Rössler (N. Y. Acad. Sci., New York 1979) p. 433
1.32 D. A. McQuarrie: Stochastic approach to chemical kinetics. J. Appl. Prob. **4**, 413 (1967)
1.33 R. Kubo, K. Matsuo, K. Kitahara: Fluctuation and relaxation of macrovariables. J. Stat. Phys. **9**, 51 (1973)
1.34 J. T. Hynes, R. Kapral, M. Weinberg: Microscopic theory of Brownian motion: Nonlinear Langevin equations. Physica **81 A**, 485 (1975)
1.35 V. Daniel: Physical Principles in Human Cooperation, The Sociological Rev. **44**, 107 (1952)
1.36 W. Weidlich: The statistical description of polarization phenomena in society. Br. J. Math. Stat. Psychol. **24**, 251 (1971)
1.37 W. Weidlich: The use of statistical models in sociology. Collect. Phenom. **1**, 51 (1972)
1.38 W: Weidlich: "Dynamics of Interacting Social Groups", in [Ref. 1.2, p. 269]
1.39 E. Callen, M. Scully, D. Shapero: J. Math. Psychol. (1975)
1.40 D. F. Walls: Non-equilibrium phase transition in sociology. Collect. Phenom. **2**, 125 (1976)
1.41 W. Weidlich, G. Haag: "Dynamics of Interacting Groups in Society with Application to the Migration of Population", in [Ref. 1.8, p. 235]
1.42 W. Weidlich, G. Haag: Migration behavior of mixed population in a town. Collect. Phenom. **3**, 89 (1980)
1.43 J. S. Coleman: *Introduction to Mathematical Sociology,* (The Free Press, New York 1964)
1.44 D. J. Bartholomew: *Stochastic Models for Social Processes* (Wiley, London 1967)
1.45 E. W. Montroll, W. W. Badger: *Introduction to Quantitative Aspects of Social Phenomena* (Gordon and Breach, New York 1974)
1.46 A. Nitschke: Revolutionen in Naturwissenschaft und Gesellschaft, Problemata 83, Stuttgart-Bad Cannstatt (1979)
1.47 A. Nitschke: "Systems and Changes of Systems in History", in [Ref. 1.7, p. 234]
1.48 T. Poston, I. N. Stewart, *Catastrophe Theory and Its Applications* (Pitman, London 1978)
1.49 R. Abraham, C. Shaw: *Dynamics, The Geometry of Behaviour* Vol. 0–3 (Aerial Press, Santa Cruz 1982)
1.50 M. W. Hirsch, S. Smale: *Differential Equations, Dynamical Systems and Linear Algebra* (Academic, New York 1974)
1.51 R. L. Stratonovich: *Topics in the Theory of Random Noise,* Vols. 1 and 2 (Gordon and Breach, New York 1963 and 1967)
1.52 A. T. Bharucha-Reid: *Elements of the Theory of Markov Processes and Their Applications* (Mc-Graw Hill, New York 1960)
1.53 N. Wax (ed.): *Selected Papers on Noise and Stochastic Processes* (Dover, New York 1954)

1.54 D. H. Sattinger: "Topics in Stability and Bifurcation Theory", Lecture Notes Math., Vol. 309 (Springer, Berlin, Heidelberg, New York 1973)
1.55 J. E. Marsden, M. Mc Cracken: *The Hopf Bifurcation and Its Applications,* Appl. Math. Sc. Vol. 19 (Springer, New York 1976)

Chapter 2

2.1 N. G. van Kampen: Can. J. Phys. **39**, 551 (1961) and in *Advances in Chemical Physics,* vol. 34, ed. by I. Prigogine, S. A. Rice (Wiley, New York 1976) p. 245
2.2 F. Haake: Decay of Unstable States, Phys. Rev. Lett. **41**, 1685 (1978)
2.3 F. Haake, J. W. Haus, R. Glauber: Passage-time statistics for the decay of unstable equilibrium states. Phys. Rev. **A 23**, 3255 (1981)
2.4 M. Suzuki: Fluctuation enhancement theory. J. Stat. Phys. **16**, 11, 477 (1977); Progr. Theor. Phys. Suppl. **64**, 402 (1978)
2.5 H. A. Kramers: Brownian motion in a field of force and the diffusion model of chemical reactions. Physica **7**, 284 (1940)
2.6 K. Matsuo, K. Lindenberg, K. E. Shuler: Relaxation time from a metastable state. J. Stat. Phys. **19**, 65 (1978)

Chapter 3

3.1 W. Weidlich, G. Haag: Quasiadiabatic solutions of Fokker-Planck equations with time-dependent drift- and fluctuation-coefficients. Z. Phys. **B 39**, 81 (1980)
3.2 H. Grabert, M. S. Green, R. Graham: Fluctuations and nonlinear irreversible processes I and II. Phys. Rev. **A 19**, 1747 (1979); **A 21**, 2136 (1980)
3.3 R. Graham, H. Haken: Generalized thermodynamic potential for Markoff systems in detailed balance and far from thermal equilibrium. Z. Phys. **248**, 289 (1971)
3.4 H. Risken: Solutions of the Fokker-Planck equation in detailed balance. Z. Phys. **251**, 231 (1972)
3.5 R. L. Stratonovich: *Topics in the Theory of Random Noise*, Vols. 1 and 2 (Gordon and Breach, New York 1963 and 1967)
3.6 H. Grabert, W. Weidlich: Renormalization of transport equations in Fokker-Planck models. Phys. Rev. **A 21**, 2147 (1980)
3.7 W. Weidlich, H. Grabert: Renormalized transport equations for the bistable potential model. Z. Phys. **B 36**, 283 (1980)
3.8 W. Weidlich: *Thermodynamik und Statistische Mechanik* (Akademische Verlagsges., Wiesbaden 1976)
3.9 K. Huang: *Statistical Mechanics* (Wiley, New York 1963)
3.10 R. Kubo: *Statistical Mechanics* (North-Holland, Amsterdam 1967)
3.11 M. Toda, R. Kubo, N. Saito: *Statistical Physics I,* Springer Ser. Solid-State Sci., Vol. 30 (Springer, Berlin, Heidelberg, New York 1983)
R. Kubo, M. Toda, N. Hashitsume: *Statistical Physics II,* Springer Ser. Solid-State Sci., Vol. 31 (Springer, Berlin, Heidelberg, New York 1983)

Chapter 4

4.1 R. M. May: *Stability and Complexity in Model Ecosystems* (University Press, Princeton 1973)
4.2 R. Rosen: *Dynamical System Theory in Biology,* Vol. 1 (Wiley-Interscience, New York 1970)
4.3 E. C. Pielou: *An Introduction to Mathematical Ecology* (Wiley-Interscience, New York 1969)
4.4 D. Ludwig: *Stochastic Population Theories,* Lecture Notes Biomath., Vol. 3 (Springer, Berlin, Heidelberg, New York 1974)
4.5 N. S. Goel, N. Richter-Dyn: *Stochastic Models in Biology* (Academic, New York 1974)
4.6 N. S. Goel, S. C. Maitra, E. W. Montroll: On the Volterra and other nonlinear models of interacting populations. Rev. Mod. Phys. **43**, 231 (1971)

4.7 E. Renshaw: Birth, death and migration processes. Biometrica **59**, 49 (1972)
4.8 W. M. Getz: Stochastic Equivalents of the linear and Volterra-Lotka systems of equations – a general birth- and death-process formulation. Math. Biosci. **29**, 235 (1976)
4.9 J. Jorne, S. Carmi: Liapunov stability of the diffusive Lotka-Volterra equations. Math. Biosci. **37**, 51 (1977)
4.10 W. Weidlich, G. Haag: Migration behaviour of mixed population in a Town. Collect. Phenom. **3**, 89 (1980)
4.11 W. Weidlich, G. Haag: „Dynamics of Interacting Groups with Application to the Migration of Population", in Lectures and Tutorials, Vol. 11 (Oldenburg, Munich 1980)
4.12 H. Haken: *Synergetics, An Introduction,* 2nd ed., Springer Ser. Synergetics, Vol. 1 (Springer, Berlin, Heidelberg, New York 1977)
4.13 W. Weidlich: On the structure of exact solutions of discrete master equations. Z. Phys. **B 30**, 345 (1978)
4.14 G. Kirchhoff: Über die Auflösung der Gleichungen auf welche man bei der Untersuchung der linearen Verteilung galvanischer Ströme geführt wird. Poggendorffs Ann. Phys. **72**, 495 (1844)
4.15 D. L. De Angelis: Application of stochastic models to a wildlife population. Math. Biosci. **31**, 227 (1976)
4.16 E. O. Wilson, W. H. Bossert: *A Primer of Population Biology* (Stamford, CT 1971)
4.17 H. C. Hanson: The Giant Canada Goose, see [4.15]
4.18 G. Haag, P. Hänggi: Exact solution of discrete master equations in terms of continued fractions. Z. Phys. **B 34**, 411 (1979)
4.19 G. Haag, P. Hänggi: Continued fraction solutions of discrete master equations not obeying detailed balance II. Z. Phys. **B 39**, 269 (1980)
4.20 W. G. Doubleday: On linear birth-death processes with multiple birth. Math. Biosci. **17**, 43 (1973)
4.21 A. J. Lotka: Analytical note on certain rhythmic relations in organic systems. Proc. Nat. Acad. Sci. U.S.A. **6**, 410 (1920)
4.22 V. Volterra: Variations and Fluctuations of the Numbers of Individuals in Coexisting Animal Populations (in italian), Mem. R. Comitato talassogr. ital., mem., 131 (1927), reprinted in Opere Math. **5**, Rome (1962)
4.23 R. Arditi, J. M. A. Billon, J. V. Da Silva: The effect of a time-delay in a predator-prey model. Math. Biosc. **33**, 107 (1977)
4.24 J. Gomatam, N. Mac Donald: Time delays and stability of two competing species. Math. Biosci. **24**, 247 (1975)
4.25 N. Mac Donald: Time delay in prey-predator models. Math. Biosci. **28**, 321 (1976)
4.26 M. Gatto, S. Rinaldi: Stability analysis of predator-prey models via the Liapunov method. Bull. Math. Biol. **39**, 339 (1977)
4.27 G. F. Gause: Experimental demonstration of Volterra's periodic oscillations in numbers of animals. J. Exp. Biol. **12**, 44 (1934)
4.28 C. B. Huffaker: Experimental studies on predation: Dispersion factors and predator-prey oscillations. Hilgardia **27**, 343 (1958)
4.29 S. Utida: On the equilibrium state of the interacting population of an insect and its parasite. Ecology **31**, 165 (1950)
4.30 H. Yoon, H. W. Blanch: The stability of predator-prey interactions in microbial ecosystems. Math. Biosci. **33**, 85 (1977)
4.31 W. C. Chewning: Migratory effects in predator-prey models. Math. Biosci. **23**, 253 (1975)

Chapter 5

5.1 J. M. Keynes: "The General Theory and After, Part II: Defence and Development" in: D. Moggeridge, (ed.) Keynes (Collective Writings, Vol. 14, London 121, 1973)
5.2 J. R. Hicks: *A Contribution to the Theory of the Business Cycles,* 3rd ed., Clarendon Press, Oxford 1956)
5.3 M. Kalecki: A macrodynamic theory of business cycles. Econometrica **3**, 327 (1935)
5.4 E. W. Montroll, W. W. Badger: *Introduction to Quantitative Aspects of Social Phenomena* (Gordon and Breach, New York 1974)

5.5 R. M. Goodwin: A non-linear theory of the cycle. Rev. Economics and Statistics **32**, (1950)
5.6 W. Chang, D. Smyth: The existence and persistence of cycles in a nonlinear model, Kaldor's 1940 model re-examined. Rev. Economics and Statistics **32**, (1950)
5.7 J. M. Clark: Business acceleration and the law of demand. J. Political Economy XXV, 217 (1917)
5.8 P. A. Samuelson: Interaction between the multiplier analysis and the principle of acceleration. Rev. Economics and Statistics **21**, 75 (1939)
5.9 J. Kornai: *Anti-Äquilibrium* (Springer, Berlin, Heidelberg, New York 1974)
5.10 E. Slutzky: The summation of random causes as the source of cyclic processes, problems of economic conditions. Econometrica **5**, 105 (1937)
5.11 G. Mensch, G. Haag, W. Weidlich: The Schumpeter clock. Econometrica, forthcoming (1982)
5.12 R. M. Goodwin: The nonlinear accelerator and the persistence of business cycles. Econometrica **19**, 1 (1951)
5.13 J. A. Schumpeter: *Konjunkturzyklen, eine theoretische, historische und statistische Analyse des kapitalistischen Prozesses*, Vol. 2 (Vandenhoeck und Ruprecht, Göttingen 1961)
5.14 G. Mensch: *Stalemate in Technology* (Ballinger, Cambridge, Ma. 1979)
5.15 Y. Barzel: Optimal timing of innovations. Rev. Economics and Statistics **50**, 348 (1968)
5.16 M. L. Kamien, N. L. Schwartz: Timing of innovation under rivalry. Econometrica **40**, 43 (1972)
5.17 N. Rosenberg: On technological expectations. Economic J. **86**, 523 (1976)
5.18 O. Glaser, A. L. Strauss: Awareness contexts and social interaction. Am. Sociological Rev. **29**, 669 (1964)
5.19 M. T. Flaherty: Industry structure and cost-reducing investment. Econometrica **48**, 1187 (1980)
5.20 W. E. G. Salter: *Productivity and Technical Change* (Ballinger, Cambridge, MA 1960)
5.21 G. Mensch, K. Kaasch, A. Kleinknecht, R. Schnopp: Innovation, Investment Trends and Switches Between Full- and Underemployment Equilibria, Berlin: International Institute of Management, Discussion Paper 10 (1980)
5.22 M. S. Feldstein, D. K. Foot: The other half of gross investment: Replacement and modernization expenditures. Rev. Economics and Statistics **53**, 49 (1971)
5.23 D. W. Jorgenson: Capital theory and investment behaviour. Am. Economic Rev. **53**, 241 (1963)
5.24 D. W. Jorgenson, J. A. Stephenson: Investment behaviour in U.S. manufacturing: 1947–1960. Econometrica **35**, 169 (1967)
5.25 R. Eisner: *Factors in Business Investment*, National Bureau of Economic Research General Series No. 102 (Ballinger, Cambridge, MA 1978)
5.26 J. M. Utterback, W. J. Abernathy: A dynamic model of process and product innovation, Omega, Intern. J. Management Sci. **3**, 639 (1975)
5.27 W. Weidlich, G. Haag: An Indicator of Change in the Investment Structure, to appear in Proceedings of the Berlin Conference on "Modelling Innovation Processes and Structural Change" ed. by G. O. Mensch (International Institute of Management, Science Center, Berlin 1982)

Chapter 6

6.1 D. R. Hofstadter: *Gödel, Escher, Bach, an Eternal Golden Braid* (Penguin Books, New York 1980)
6.2 D. H. Meadows, et al.: *The Limits to Growth (A Report of the Club of Rome's Project on the Predicament of Mankind)*, 2nd ed. (Universe 1974)
6.3 D. Meadows et al.: *Alternatives to Growth* (Ballinger, Cambridge, MA 1977)
6.4 D. Meadows et al.: *Dynamics of Growth in a Finite World* (MIT Press, Cambridge, MA 1974)
6.5 D. Meadows et al.: *Toward Global Equilibrium* (MIT Press, Cambridge, MA 1973)
6.6 K. W. Deutsch: *Politische Kybernetik, Modelle und Perspektiven* Vol. 3 (Verlag Rombach 1973)
6.7 J. W. Forrester: *World Dynamics* (Wright-Allen Press 1971)
6.8 T. Poston, I. N. Stewart: *Catastrophe Theory and Its Applications* (Pitman, London 1978)

Subject Index

Springer Series in
Synergetics

Series Editor: **H. Haken**

Volume 1: **H. Haken**
Synergetics
An Introduction. Nonequilibrium Phase Transitions and Self-Organization in Physics, Chemistry and Biology. 2nd enlarged edition. ISBN 3-540-08866-0

Volume 2:
Synergetics
A Workshop. Proceedings of the International Workshop on Synergetics at Schloß Elmau, Bavaria, May 2–7, 1977
Editor: **H. Haken**
ISBN 3-540-08483-5

Volume 3:
Synergetics
Far from Equilibrium. Proceedings of the Conference Far From Equilibrium: Instabilities and Structures, Bordeaux, France, September 27–29, 1978
Editors: **A. Pacault, C. Vidal**
ISBN 3-540-09304-4

Volume 4:
Structural Stability in Physics
Proceedings of Two International Symposia on Applications of Catastrophe Theory and Topological Concepts in Physics, Tübingen, Federal Republic of Germany, May 2–6 and December 11–14, 1978
Editors: **W. Güttinger, H. Eikemeier**
ISBN 3-540-09463-6

Volume 5:
Pattern Formation by Dynamic Systems and Pattern Recognition
Proceedings of the International Symposium on Synergetics at Schloß Elmau, Bavaria, April 30–May 5, 1979
Editor: **H. Haken**
ISBN 3-540-09770-8

Volume 6:
Dynamics of Synergetic Systems
Proceedings of the International Symposium on Synergetics, Bielefeld, Federal Republic of Germany, September 24–29, 1979
Editor: **H. Haken**
ISBN 3-540-09918-2

Volume 7: **L. A. Blumenfeld**
Problems of Biological Physics
ISBN 3-540-10401-1

Volume 8:
Stochastic Nonlinear Systems
in Physics, Chemistry and Biology. Proceedings of the Workshop, Bielefeld, Federal Republic of Germany, October 5–11, 1980
Editors: **L. Arnold, R. Lefever**
ISBN 3-540-10713-4

Volume 9:
Numerical Methods in the Study of Critical Phenomena
Proceedings of a Colloquium, Carry-le-Rouet, France June 2–4, 1980
Editors: **J. Della Dora, J. Demongeot, B. Lacolle**
ISBN 3-540-11009-7

Volume 10: **Y. L. Klimontovich**
The Kinetic Theory of Electromagnetic Processes
ISBN 3-540-11458-0

Volume 11:
Chaos and Order in Nature
Proceedings of the International Symposium on Synergetics at Schloß Elmau, Bavaria, April 27 – May 2, 1981
Editor: **H. Haken**
ISBN 3-540-11101-8

Volume 12:
Nonlinear Phenomena in Chemical Dynamics
Proceedings of an International Conference, Bordeaux, France, September 7–11, 1981
Editors: **C. Vidal, A. Pacault**
ISBN 3-540-11294-4

Volume 13: **C. W. Gardiner**
Handbook of Stochastic Methods
for Physics, Chemistry and the Natural Sciences
ISBN 3-540-11357-6

Volume 14: **W. Weidlich, G. Haag**
Concepts and Models of a Quantitative Sociology
The Dynamics of Interacting Populations
ISBN 3-540-11358-4

Volume 15: **W. Horsthemke, R. Lefever**
Nonequilibrium Transitions Induced by External Noise
ISBN 3-540-11359-2

Volume 16: **L. A. Blumenfeld**
Physics of Bioenergetic Processes
ISBN 3-540-11417-3

Volume 17:
Evolution of Order and Chaos
in Physics, Chemistry, and Biology
Proceedings of the International Symposium on Synergetics at Schloß Elmau, Bavaria, April 26 – May 1, 1982
Editor: **H. Haken**
ISBN 3-540-11904-3

Springer-Verlag
Berlin
Heidelberg
New York

MIX
Papier aus verantwortungsvollen Quellen
Paper from responsible sources
FSC® C105338

If you have any concerns about our products,
you can contact us on
ProductSafety@springernature.com

In case Publisher is established outside the EU,
the EU authorized representative is:
Springer Nature Customer Service Center GmbH
Europaplatz 3, 69115 Heidelberg, Germany

Printed by Libri Plureos GmbH
in Hamburg, Germany